Emerging Trends in Mechatronics

Edited by Aydin Azizi

Published in London, United Kingdom

IntechOpen

Supporting open minds since 2005

Emerging Trends in Mechatronics
http://dx.doi.org/10.5772/intechopen.81944
Edited by Aydin Azizi

Contributors
Mohammadreza Koopialipoor, Amin Noorbakhsh, Shahin Zareie, Abolghassem Zabihollah, Izzat Al-Darraji, Ali Kılıç, Sadettin Kapucu, Yan Ran, Shengyong Zhang, Genbao Zhang, Xinlong Li, Erika Ottaviano, Pierluigi Rea, Ľuboslav Straka, Gabriel Dittrich, Valery Kokovin, Kanstantsin Miatliuk, Lefteris Katrantzis, Vassilis Moulianitis, M. Shahria Alam

First published in London, United Kingdom, 2020 by IntechOpen
IntechOpen is the global imprint of INTECHOPEN LIMITED, registered in England and Wales, registration number: 11086078, 7th floor, 10 Lower Thames Street, London, EC3R 6AF, United Kingdom
Printed in Croatia

British Library Cataloguing-in-Publication Data
A catalogue record for this book is available from the British Library

Additional hard and PDF copies can be obtained from orders@intechopen.com

Emerging Trends in Mechatronics
Edited by Aydin Azizi
p. cm.
Print ISBN 978-1-78984-319-4
Online ISBN 978-1-78984-320-0
eBook (PDF) ISBN 978-1-83968-505-7

An electronic version of this book is freely available, thanks to the support of libraries working with Knowledge Unlatched. KU is a collaborative initiative designed to make high quality books Open Access for the public good. More information about the initiative and links to the Open Access version can be found at www.knowledgeunlatched.org

For EU product safety concerns:
IN TECH d.o.o., Prolaz Marije Krucifikse Kozulić 3, 51000 Rijeka, Croatia, info@intechopen.com or visit our website at intechopen.com.

Meet the editor

Prof. Aydin Azizi holds a PhD degree in Mechanical Engineering and currently serves as an assistant professor and research focal point of the Research Council of Oman at the German University of Technology in Oman where he conducts the Siemens Mechatronic Systems Certification Program as an official Siemens-certified mechatronics instructor. His current research focuses on investigating and developing novel techniques to model, control, and optimize complex systems. Prof. Azizi's areas of expertise include control and automation, artificial intelligence (AI), and simulation techniques. Prof. Azizi is the recipient of the National Research Award of Oman for his AI-focused research, DELL EMC's "Envision the Future" completion award in IoT for "Automated Irrigation System," and "Exceptional Talent" recognition by the British Royal Academy of Engineering.

Contents

Preface

Mechatronics is the combination of mechanical, electrical and electronics, control and automation, and computer engineering. The main research task of mechatronics is the design, control, and optimization of advanced devices, products, and hybrid systems utilizing the concepts found in all these fields. In general, the purpose of this special issue is to help better understand how mechatronics will impact on the practice and research of developing advanced techniques to model, control, and optimize complex systems. The special issue presents recent advances in mechatronics and related technologies, including: automatic control, robotics, agent-based systems, smart manufacturing, and Industry 4.0. The selected topics give an overview of the state of the art and present new research results and prospects for the future development of the interdisciplinary field of mechatronic systems. This special issue provides up-to-date and useful knowledge for researchers and engineers involved in mechatronics and related fields.

Aydin Azizi
German University of Technology in Oman

Chapter 1

Intelligent Control System of Generated Electrical Pulses at Discharge Machining

Ľuboslav Straka and Gabriel Dittrich

Abstract

The book chapter provides a comprehensive set of knowledge in the field of intelligent control of generated electrical impulses for wire electrical discharge machining. With the designed intelligent electrical pulse control system, the stability of the electroerosion process, as well as the increased surface quality after wire electrical discharge machining (WEDM), can be significantly enhanced compared to standard impulse control systems. The aim of the book chapter is also to point out the importance of monitoring in addition to the established power characteristics of generated electrical pulses, such as voltage and current, as well as other performance parameters. The research was mainly focused on those parameters that have a significant impact on the quality of the machined surface. The own's theoretical and knowledge base was designed to enrich the new approach in increasing the geometric accuracy of the machined surface, as well as the overall efficiency of the electroerosion process for WEDM through intelligent control of generated electrical pulses.

Keywords: adaptive system, acoustic emission, automation, control system, discharge machining, pulse generator, spark, quality

1. Introduction

The current trend in the development of mechanical engineering carries signs of complexity and dynamism. At the same time, it is increasingly influenced by new scientific and technical knowledge and requirements for their rapid deployment. For the production of high-precision components of state-of-the-art and highly-sophisticated technical equipment, fully automated production systems and progressive manufacturing technologies are often used. In most cases, an integral part of them is a management system that manages demanding technological processes. Application of the given system provides a suitable precondition for ensuring the required high quality of manufactured products.

Another, not less significant trend at present is the focus on the development of scientific and technical principles of modern engineering production. At the same time, links with classical teachings are being sought, with emphasis on their direct use in technical practice. This trend is also aided by statistical, optimization and simulation methods. These have been used only in the past in the field of mechanical engineering for the solution of partial technological tasks. For example, they allowed a basic selection of technological process variants.

The current rapid development of computer technology creates wide scope for the use of mathematical methods in both theoretical and practical technological tasks. Cybernetic methods, probabilistic logic, mathematical modeling and simulation of production processes are used in connection with the development of computer technologies. All of this increases the demands on the degree of exactness of the formulation of knowledge, as well as the efficiency and quality of technological solutions, which aim to save the work of engineers, technicians and workers.

In addition, the continuous development of modern mechanical engineering places increased demands on the introduction of advanced production methods, advanced production facilities and their control systems. Particular attention is paid to machining processes in which, in particular, the mechanical properties of the workpiece and the tool do not impose almost any limits. These are, in particular, machining methods in which the degree of machinability of a material is dependent only on physical properties such as e.g. thermal and electrical conductivity, melting temperature, atomic valence and the like. As already mentioned, their essential part is computer support. A computer-aided production process has a huge advantage in that the human factor of poor product quality is almost excluded. In this case, the quality of the machined surface depends directly on the design of the machine, its software management and the setting of technological and process parameters.

Undoubtedly these processes include WEDM, where the decisive link with the primary impact on the quality of the machined surface is the electrical pulse generator. Nowadays, various types of electrical pulse generators are used for WEDM, the vast majority of which control performance parameters to maximize performance. It is exactly the new type of generator of electrical impulses applicable in the conditions of the electroerosion process which is described by Qudeiri et al. [1]. In the control algorithm of a given type of electric pulse generator, there is absolutely no criterion relating to the geometric accuracy of the machined surface. Researchers Yan and Lin [2] in turn dealt with the development of a new type of pulse generator which, unlike the previous type, is not oriented to maximize performance, but minimize the surface roughness of the machined surface. A similar type of pulse generator is also described by Świercz and Świercz [3]. However, even in this case, there is no qualitative criterion for the geometric accuracy of the machined surface. Researchers Barik and Rao [4] participated in the development of a special type of electrical pulse generator designed for electrical discharge machining in laboratory conditions. Although their newly developed generator allows to set the operating parameters of the electric pulse generator according to the specific quality requirements of the machined surface, the criterion of geometric accuracy of the machined surface is missing again.

Thus, it is clear from the above overview that insufficient attention is paid to the development of electrical pulse generators with a focus on the geometric accuracy of the machined surface. Therefore, the aim of this chapter of the book is to contribute to the database of existing knowledge in the field of intelligent system design for precise control of generated electrical pulses for WEDM with the focus on maximizing the geometric accuracy of the machined surface. These findings are intended to help improve the quality of components produced by the progressive WEDM technology, the practical application of which is described in detail in Chapter 2. This is based on the physical nature of the material removal described in detail in Chapter 3. Chapter 4 deals with the current state of electrical discharge parameter control during WEDM, which highlights the current deficiencies of current approaches in the control of generated electrical pulses. Chapter 5 describes possible approaches to eliminate tool electrode vibration during WEDM, by applying measures regarding technological and process parameters. It also points to the application of one of the acceptable options that concerns the innovation of an intelligent control system for generated electrical pulses during the electroerosion process. Further, based on an analysis of current modern approaches in the construction of electrical pulse generators used for WEDM, detailed in Chapter 6, an adaptive control system for generated electrical pulses was designed during WEDM. This innovated control system for generated electrical pulses, designed to increase the geometric accuracy of the machined surface for WEDM, is described in detail in the Chapter 7.

2. Application of progressive technologies in technical practice

As already mentioned in the introduction, modern engineering production currently places high demands on the mechanical properties of the materials used. The emphasis is mainly on their high strength, hardness and toughness. Therefore, materials such as various types of high-strength and heat-resistant alloys, carbides, fiber-reinforced composite materials, stelites, ceramic materials and advanced composite tooling materials, etc., are at the forefront. At the same time, with the use of these high-strength materials, the demands on accuracy and also on the performance of machine tools and equipment increase. These facts necessitate the development and deployment of machining methods that allow high material removal while achieving high quality machined surfaces. In this respect, there are some advantages to those machining methods in which there is no mechanical separation of the material particles. The application of these progressive machining methods to technical practice is particularly accentuated by the fact that not only the mechanical properties of the material, but also other properties such as thermal and electrical conductivity, melting temperature, atomic valency, density and the like, determine the machinability limits. Another not less important reason for implementing progressive machining methods is the complicated geometrical shapes of the workpiece, which often require demanding manufacturing processes. This results in long machining times, the use of special tools, special fixtures and the like. These are usually very expensive. A perfect control system is needed to meet all the above requirements. Standard processes for managing production processes are already inadequate today. Especially those who can adequately adapt to the current situation and the needs of the machining process are entering the forefront.

One of the progressively developing technologies in the field of machining process management is electrical discharge machining (EDM). Moreover, the essence of the production of components with the application of this progressive technology is based on the fact that the mechanical properties of the machined material do not impose almost any limits on its machining. The only limiting factor for the machinability of these materials is their appropriate chemical and physical properties. This technology is principally based on the use of thermal energy to which the electrical discharge generated between the two electrodes is transformed, of which the first electrode represents the tool and the second workpiece. It is a machining process in which material removal occurs through cyclically repeated electrical discharges. Through these, the microscopic particles in the form of beads are removed from the material by melting and subsequent evaporation in conjunction with high local temperature. It moves at a level 10,000°C. However, the electroerosion process must be precisely controlled by a reliable control system.

3. Physical nature of material removal for WEDM

To ensure precise management of the electroerosion process during WEDM, it is essential to base it on its physical principle. The physical principle of material removal for WEDM can be regarded as a relatively challenging and complicated process. Its essence lies in the formation of a discharge between two electrodes (tool—workpiece) either in very thin gas, in air, or in gas at normal temperature and pressure, or in a dielectric fluid, i. e. in a fluid with high electrical resistance. However, the classical electrical discharges that occur between the two electrodes (tool—workpiece) in the gas dielectric have relatively little effect. Therefore, such an environment is not quite ideal for the needs of precision and high-performance machining. In this regard, the application of fluid dielectric media is much more advantageous. These dielectrics significantly increase the effect of electrical discharges between the electrodes (tool—workpiece). Electrically charged particles, electrons and ions are the active agents in the erosion of material particles from the surface of both electrodes. They are formed as a product in the ionization process. Subsequently, in the electric field, they acquire the kinetic energy that, along with the output work, is passed on the surface of both electrodes. The shape and size of the eroded metal particles from the material being machined, as well as the size and shape of the resulting crater (**Figure 1**) depend not only on the polarity of the electrodes, but also on the particular application of the technological parameters.

By default, 10^{-3} až 10^{-5} mm^3 of material is removed by WEDM during a single discharge cycle by electroerosion. Its size can be empirically determined by the relationship (1):

$$V_i = K \cdot W_i \tag{1}$$

where, V_i (mm^3) is the volume of material taken, K ($mm^3.J^{-1}$) is the proportionality factor for cathode and anode, W_i (J) is the discharge energy.

As mentioned above, the shape and size of the crater formed in both electrodes during one discharge cycle depends mainly on the magnitude of the applied discharge energy. This is given by the specific setting of technological parameters. The time course of individual discharges is characterized by several indicators. These are indicators relating to the discharge current I (A), the discharge voltage U (V) and the duration of the individual discharges t_{on} (μs), as well as the breaks t_{off} (μs) between discharges. The events that take place between the two electrodes during the electroerosion process are comprehensively described the volt-ampere characteristic. This is shown in **Figure 2**.

The total volume of V_T, material taken from both electrodes during the electroerosion process is directly dependent on the magnitude of the transmitted

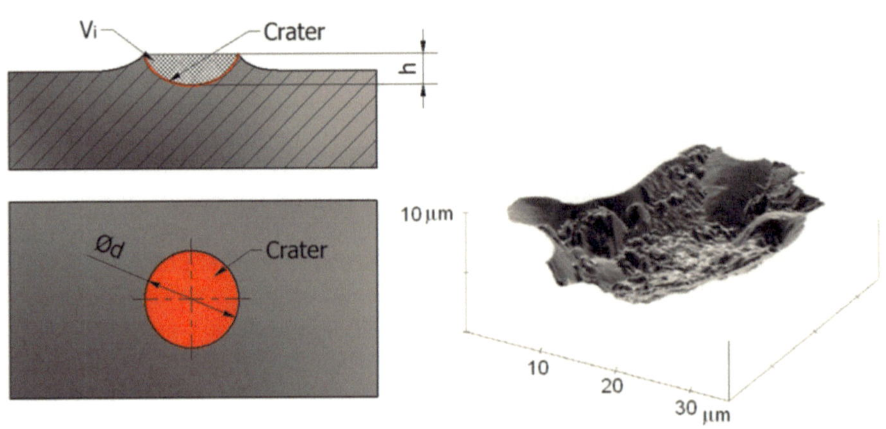

Figure 1.
The shape and size of the crater formed during one discharge cycle. V_i—volume of material taken, h—depth of crater, d—crater diameter.

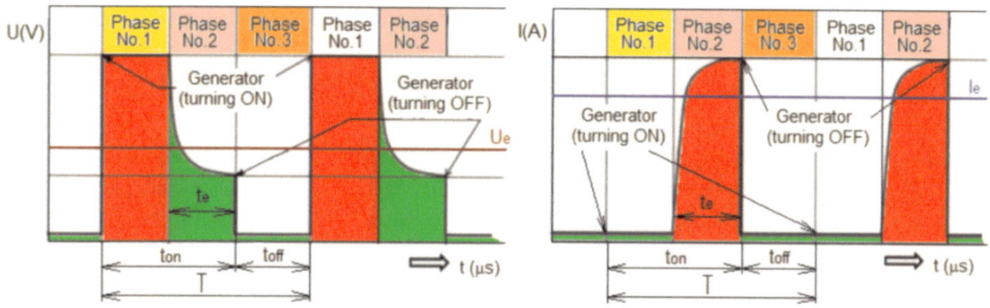

Figure 2.
Volt-ampere characteristic of one discharge cycle during the electroerosion process.

energy W_e. This in turn results in a series of cyclically repeating electrical discharges between the electrodes (tool—workpiece) over time t. The total discharge energy W_e transmitted during a series of discharge cycles can be empirically determined by the relationship (2):

$$W_e = \int_0^T U(t) \cdot I(t) \mathrm{d}t \tag{2}$$

where, W_e (J) is the total discharge energy, $U(t)$ (V) is the electrode discharge voltage at time t, $I(t)$ (A) is the maximum discharge current at time t, T (μs) is the duration of one period of electrical discharge.

By deriving the relation (2), the amount of energy transmitted during one discharge cycle can then be empirically determined (3):

$$W_e = I_e \cdot U_e \cdot t_{on} \tag{3}$$

where, I_e (A) is the average discharge current, U_e (V) is the average discharge voltage on the electrodes, t_{on} (μs) is the duration of discharge during one discharge cycle (delayed generator operation).

In order to complete all the parameters of the electroerosion process related to one discharge cycle, it is also necessary to empirically determine the magnitude of the average discharge current I_e and the discharge voltage U_e between the electrodes. These values can be determined based on the relationship (4) for I_e and the relation (5) for U_e:

$$I_e = \frac{1}{t_e} \int_{0_{t_e}} I(t) \mathrm{d}t \tag{4}$$

$$U_e = \frac{1}{t_e} \int_{0_{t_e}} U(t) \mathrm{d}t \tag{5}$$

where, $I(t)$ (A) is the maximum discharge current (A), t_e (μs) is the current discharge time (generator operation).

Based on these and other parameters of the electroerosion process, the total amount of material taken per time unit t can then be empirically determined by relation (6):

$$Q_T = k \cdot r \cdot f \cdot \mu \cdot W_e = k \cdot r \cdot f \cdot \mu \cdot \int_0^T U(t) \cdot I(t) \mathrm{d}t \tag{6}$$

where, Q_T ($mm^3 \cdot s^{-1}$) is the total amount of material withdrawn per time unit t, k is the factor of proportionality for cathode and anode, r is the efficiency of electrical discharge, f (s^{-1}) is the frequency of electrical discharges, μ is the efficiency of the discharge generator.

Another not less important parameter in specifying electroerosion process parameters is the discharge period t_d. This characterizes the overall efficiency of one discharge cycle between the electrodes. It is empirically determined as a proportion of the duration of the electrical discharge t_{on} during one discharge cycle, that is, the time between the generator on and off and the period of time T, that is, the time interval between two consecutive generator starts. Its value can be determined by relation (7):

$$t_d = \frac{t_{on}}{T} = \frac{t_{on}}{t_{on} + t_{off}} \tag{7}$$

where, t_d is the discharge period, t_{on} (μs) is the duration of the discharge during one discharge cycle (delayed generator operation), t_{off} (μs) is the break time between two consecutive discharges, T (μs) is the electric discharge period time.

Figure 3 describes the effect of the discharge period t_d on the machined surface quality for WEDM in terms of the roughness parameters Ra and Rz.

In addition, by using the discharge period t_d, we can empirically express the overall efficiency of one discharge cycle between the electrodes during electroerosion, we can also quantify individual types of electrical discharges. Its

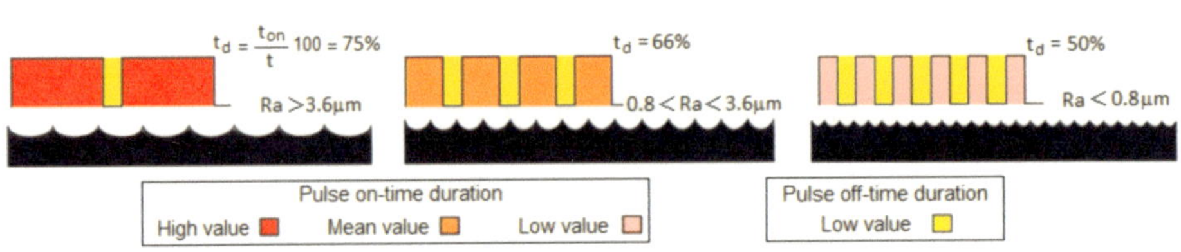

Figure 3.
Effect of discharge period td in the range of 50–75% on the machined surface quality forWEDM in terms of parameters Ra and Rz.

Electrical discharge parameters	Type of discharge	
	Electric spark	**Short term electric arc**
Total discharge duration t_i (μs)	Short time (0.1 až 10^{-2})	Long time (>0.1)
Time usage of discharge period t_d	Low value (0.03 až 0.2)	High value (0.02 až 1)
Discharge frequency	High value	Low value
Current density at the discharge point ($A.mm^{-2}$)	Approx. 10^6 $A.mm^{-2}$	10^2–10^3 $A.mm^{-2}$
The discharge channel temperature (°C)	High (over 10,000)	Low (in the range of 3300–3600)
Energy individual discharges W_e (J)	Low (10^{-5}–10^{-1})	High (approx. 10^2)
Practical use for WEDM	High quality machined surface (finishing)	Low quality of machined surface (rouhing)

Table 1.
Basic property specification of stationary and non-stationary type of discharge.

value makes it easier to identify the type of electric arc, i.e. whether it is stationary electric spark or nonstationary short term electric arc. At the same time, its value is to some extent influenced by the total amount of Q_T of the withdrawn material per time unit t, as well as the resulting quality of the machined surface.

It can be seen from **Table 1** that precise control of the individual electrical discharge parameters between the two electrodes during a single discharge cycle has a significant impact on the quality of the machined surface for WEDM as well as the overall efficiency of the electroerosion process.

4. Current state in the field of electric discharge parameter control for WEDM

From the point of view of achieving the high quality of the machined surface for WEDM as well as the high overall efficiency of the electroerosion process, the discharge process must be precisely software controlled. The current trend in the development of electrical pulse generators is mainly focused on control systems that do not measure effective voltage, but separately the duration of each discharge. These times are summed and the intensity of the electric discharge is regulated accordingly. This data is then used to control the actuator, while the response time and magnitude of the servo motion correspond to that in the working gap (**Figure 4**).

Recently, mainly used so-called alternating electric pulse generators. There are also generators of electrical impulses in which is voltage with the same polarity supplied between tool electrode and workpiece. This causes the ions to pass in one direction, causing increased corrosion of the eroded material. However, if the electrical voltage polarity is alternated with a certain frequency, this effect is suppressed and corrosion does not occur. Moreover, the practical advantage of applying these types of pulse generators is a narrower working gap. This type of generator finds its application especially in the erosion of carbide alloys.

As mentioned above, the process of controlling electrical impulses during WEDM is in some cases based on predetermined mathematical models. Important pioneers in this area were Scott et al. [5]. In particular, they focused their research on modeling performance parameters for electroerosion machining in various conditions and modes. In their research, they also found that there is no single combination of levels of important factors that can be optimal at all times. Research in the field was also addressed by Tarng et al. [6, 7]. They then formulated mathematical models that allow predicting the achieved quality of the machined surface depending on the setting of the electric discharge parameters using simulation and optimization elements. Researchers Sarkar et al. [8] have devoted a substantial part

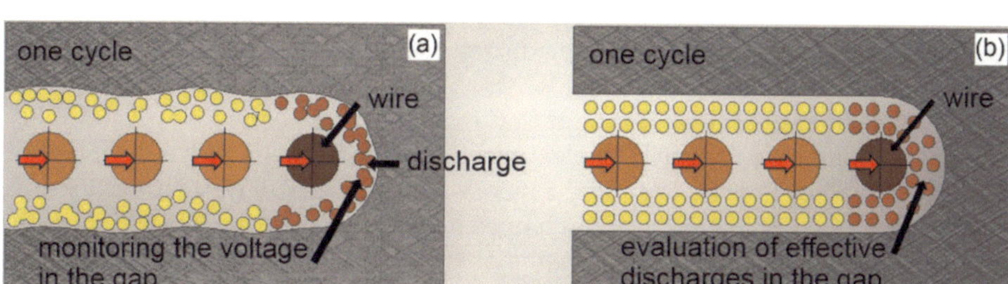

Figure 4.
Controls of actuators for WEDM based on the system for monitoring the duration of individual discharge cycles. (a) Monitoring the effective voltage in the gap and (b) evaluation of the time duration of the discharge.

of their research to the mathematical modeling of the achieved surface quality at WEDM depending on the current electrical discharge parameters using neural networks. A detailed study of the electrical discharge performance parameters during WEDM based on their mathematical modeling was done by Puri and Bhattacharyya [9]. The areas of modeling of electrical discharge parameters, considering the polarity change of electrodes during WEDM, have been researched by Liao and Yub [10]. At the same Mahapatra and Patnaik [11] they described in detail the possibilities of using the nonlinear modeling method to optimize these parameters. A specific area of electroerosion process control has been studied by Jin et al. [12]. In their research, they have developed a combined structural model that describes the use of thermal energy, including a balance of the effects of vibration on the stability of the electric arc. In addition, the model also included high temperature effects due to high-power electrical discharges. Yan et al. [13, 14] described a new approach in electrical discharge parameters optimization during electrical discharge machining based on selected performance characteristics such as maximum discharge current I (A), maximum electrical discharge voltage U (V), discharge duration during one discharge cycle t_{on} (μs) and the duration of the break between discharges t_{off} (μs). All these parameters have been optimized with regard to the quality and efficiency of the electroerosion process, as well as to minimize wear on the tool electrode. The physical aspect of the electroerosion process was addressed by Kopac [15]. In his experimental research, he tested various power parameters of the electric arc with respect to the content of electrically conductive parts in the discharge channel during the electroerosion process. He found that with the increasing share of electrically conductive parts in the discharge channel during the electroerosion process, its performance and productivity increased. At the same time, it points out that the main electrical parameters of the electric arc during the electroerosion process have the smallest influence on the crater size after the electric discharge of the maximum discharge voltage U. Only its crater shape changes with its size. The study of vibration of the tool electrode due to electrical discharges during the electroerosion process was investigated by Shahruz [16]. They found that tool electrode vibration has a significant contribution to the geometric inaccuracy of the machined surface after WEDM. They also argue in their study that the high tool electrode tension forces near critical values have a positive effect on reducing the amplitude of the wire electrode vibration during the electroerosion process, but cannot completely eliminate them. The vibration of the tool electrode was also investigated by Altpeter and Roberto [17]. In particular, their research was substantiated by the fact that the issue of damping tool electrode vibration during WEDM has been poorly addressed in the past. The shape and size of the craters after the individual electrical discharges during the WEDM were dealt with by Hewidy and Gokler [18, 19]. They tried to describe mathematically the influence of the magnitude of the discharge energy during the electroerosion process on the size and shape of the craters. They found that higher values of maximum discharge current I (A) and duration of discharge during one discharge cycle t_{on} (μs) contribute to increase in crater size.

From the above overview, it is clear that several experimental investigations have been conducted in the field of electrical discharge between the two electrodes during one discharge cycle. Despite the increasing emphasis on the complexity of learning about the set of electrical discharge characteristics during WEDM, there is still a lack of comprehensive identification of their interconnections. At the same time, there are no suggestions for minimizing the adverse effect of electrical discharges on the quality of the machined surface after WEDM in terms of geometric accuracy.

5. Elimination of tool electrode vibration for WEDM

During the duration of the individual electrical discharges, due to the precise guiding of the tool electrode, it is necessary to adequately tension it with the force *Fw* (N). It is normally selected in the range of 5–25 N. Furthermore, it is necessary to charge the tool electrode with electrical impulses, enwrap it with a dielectric fluid and because of its wear and tear it is constantly renewed. To be able to move with such a delicate and labile tool as a few tenths of a millimeter of a thin wire electrode, very precise and sensitive guide are needed. In **Figure 5**, a part of the electroerosion CNC machine can be seen, which provides accurate guidance of the wire electrode.

The tool electrode tensioning and guiding system in the CNC electroerosive equipment consists of a supply section that grips, clamps, feeds, and controls the wire Furthermore, from the working part that guides the tool electrode through the working zone, where it is washed with dielectric fluid, supplied with electric current and subsequently eroded. The lead electrode guidance system is terminated by a drain portion, which retracts the electrode, rechecks it, and wraps it onto the coil.

As already mentioned, in addition to the inaccuracy in the wiring of the wire electrode, its deflection from a straight position causes forces that arise due to the cyclic action of the electrical discharges between the two electrodes [20]. The rules applies that the greater the thickness of machined material, the greater the deflection. Partial compensation for this adverse event is carried out by special measures. In particular, the inclusion of counter force in conjunction with a system that ensures optimum tension of the tool electrode. However, neither of these measures has a sufficient effect.

In addition to the thickness of the material being machined, the choice of the optimum tensioning force also adapts to the intensity of the electrical discharge, the type of material of the workpiece, the type of tool electrode material and its diameter, the properties of the dielectric fluid, and the like. In particular, wire electrode tension compensation serves to minimize its sag in the middle due to the cyclic action of electrical discharges with varying intensity [21]. In **Figure 6**, an extreme deflection of the tool electrode during the elektroerosion process can be observed as

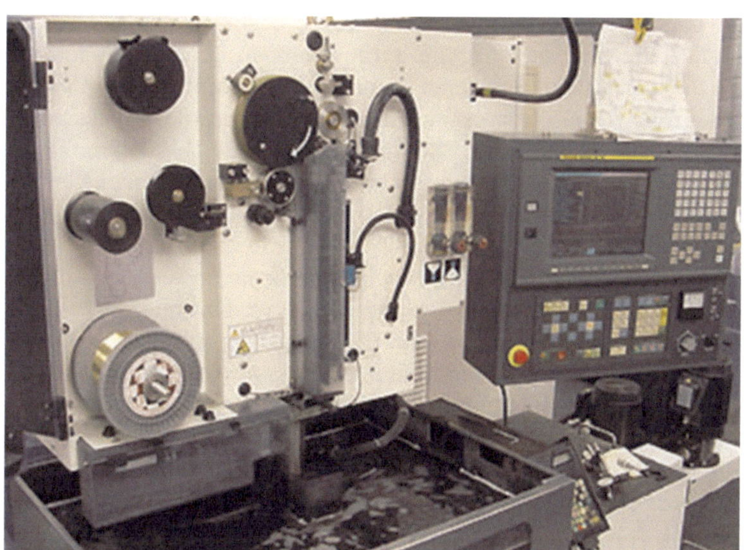

Figure 5.
Tool electrode tensioning and guiding system during the electroerosion process.

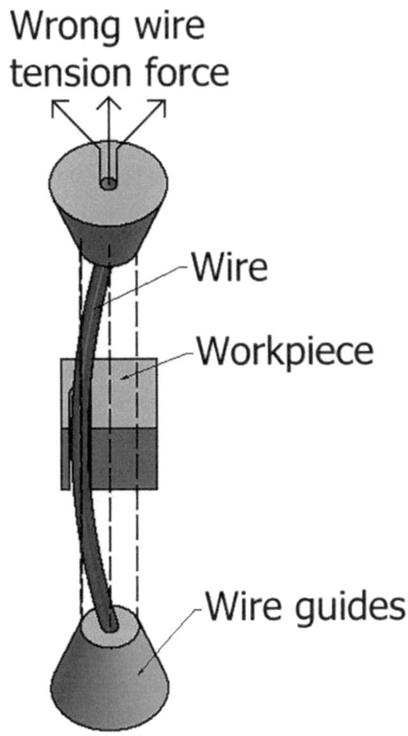

Figure 6.
Extreme deflection of the tool electrode due to the application of an improperly selected value of the compensating force during its tensioning.

a result of the inappropriate selection of the compensation force size in its stretching.

As a general rule, the higher the value of the tool electrode compensation force when it is tensioned reduces the vibration amplitude. This also leads to a reduction of the working gap, thus achieving a higher accuracy of the machined surface for WEDM. Ideally, the value of the tool electrode compensation force should be chosen to approach the material tensile strength limit [22]. However, the limit value must not be exceeded during the electroerosion process. Otherwise, the tool electrode will break. Tool electrodes with a strength in the range of 400–2000 $N.mm^{-2}$ are used as standard. Tool electrodes with a strength of up to 490 $N.mm^{-2}$ are called soft, tool electrodes with a strength of between 490 and 900 $N.mm^{-2}$ are called semi-hard and tool electrodes with a strength above 900 $N.mm^{-2}$ are called hard. **Figure** 7 shows the impact strength (hardness) of the wire tool electrode on its deflection in the electroerosion machining process when applying a constant tension force.

With increasing material thickness, it is necessary to increase the value of the tensioning force *Fw* of the tool electrode in order to eliminate vibrations. This allows, as already mentioned, a higher tensile strength value of the tool electrode material used or its increasing diameter. However, too high values of the tool electrode tension force have an adverse effect on the performance and productivity of the electroerosion process. This can be seen from the following graphical dependence on **Figure 8**.

From this graphical dependence, it can be seen that increasing the magnitude of the compensating force *Fw* when tensioning the tool electrode is in terms of productivity it has meaning only to a certain extent. When it is exceeded, there is a significant drop in the electroerosion process productivity.

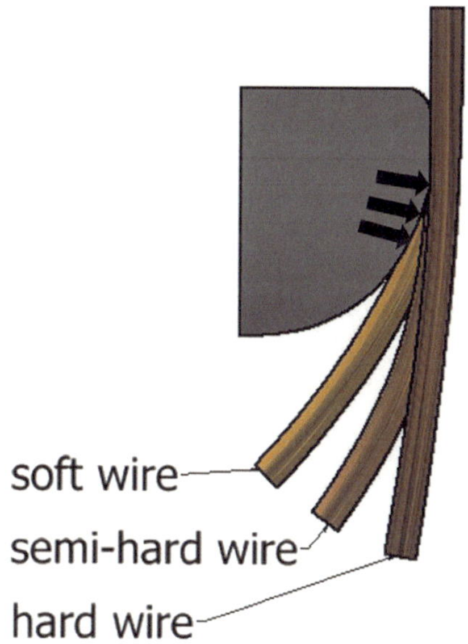

Figure 7.
Effect of hardness of wire tool electrode material on its deflection during electrical discharge machining.

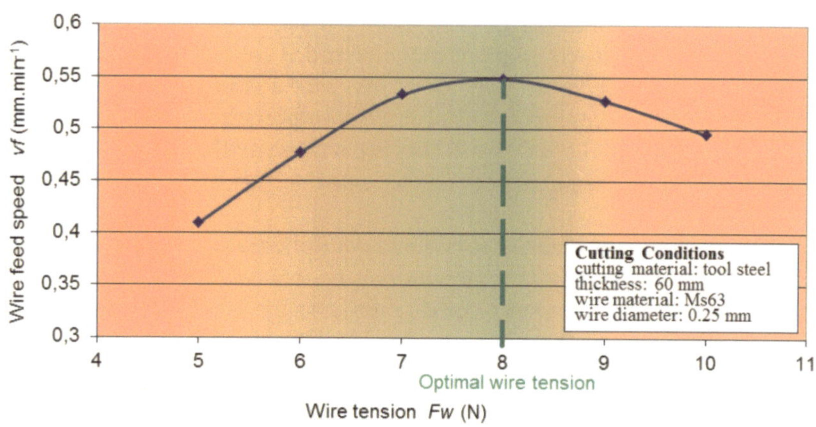

Figure 8.
The dependence of the electroerosion process productivity on the value of the compensation force Fw when the tool electrode is stretched.

Thus, it is clear from the foregoing that when applying the critical values of the compensating force *Fw* when the wire tool electrode is being tensioned, the productivity of the electroerosion process will be even lower. On the other hand, as the value increases, the vibration amplitude of the tool electrode is substantially reduced, resulting in greater geometric accuracy of the machined surface after WEDM [23]. Since the tensile strength of the wire tool electrode material is a limiting factor in the tension force selection, the choice of material type is also important [24]. By default, a single-component compact tool electrode is selected for WEDM. Materials such as Cu, Ms., Mo and the like are used. In the past, pure copper was used quite often as a tool electrode material, mainly because of its high electrical conductivity, but also in its relatively simple production. However, a significant drawback of the application of pure copper to the production of wire tool electrodes is its very low tensile strength [25]. Therefore, the Cu tool electrodes were later replaced with brass. Practical application results have shown that the presence of Zn in the tool electrode material significantly reduces the risk of breakage. This allows the application of higher values of compensation force *Fw*. Also suitable for producing tool wire electrodes is aluminum brass. This material is characterized by a tensile strength of 1200 $N.mm^{-2}$, without any adverse effect on its elongation. Although these types of tool electrode materials are less prone to damage, their usefulness in practice is relatively limited. The tooling electrodes, which are based on Mo, are used where very high tensile strength and very small wire diameter are required. In addition to the high tensile strength, this material also has a high melting point. A significant disadvantage of the application of this material is its high cost. The tungsten tool electrodes have an even greater tensile strength and a higher melting point than molybdenum. From an economic point of view, this type of material is applicable only to very small diameters ($\leq$0.05 mm) of tool electrodes [26].

As mentioned, the presence of Zn in the tool electrode material has a positive impact on its mechanical properties. However, the practical use of single-component tool electrodes with a Zn content above 40% is inefficient for economic reasons. Therefore, multi-component, for example coated electrodes have been developed for the application of higher tool electrode tensioning forces, allowing higher zinc content on the electrode surface while maintaining a homogeneous core. These tool electrodes are particularly useful when specific material requirements are required because of the high geometric accuracy of the machined surface after WEDM. In this respect, the high tensile strength of the material as well as its good electrical conductivity are decisive.

For this purpose, multi-component tool electrodes are used, the core of which is Cu, Ms. or steel and coated with pure Zn or Ms. with a zinc content of 50%. **Figure 9** shows selected combinations of multi-component tool electrodes that are used in practice for special operations. This is particularly the case when increased demands are placed on the quality of the machined surface after WEDM in terms of geometric accuracy.

These composite wires make it possible to combine traditional materials that are relatively inexpensive with expensive materials to achieve the unique properties of wire tool electrodes [27]. However, the efficiency of these coated tool electrodes is limited by the thickness of the coating which is relatively thin. The standard ranges from 5 to 10 μm. A special case consists of three-component tool electrodes, whose core is a steel wire. It is coated with a layer of copper and brass with 50% zinc content. The coated tool electrodes allow the application of relatively high tension forces *Fw* while maintaining an acceptable electroerosion process productivity.

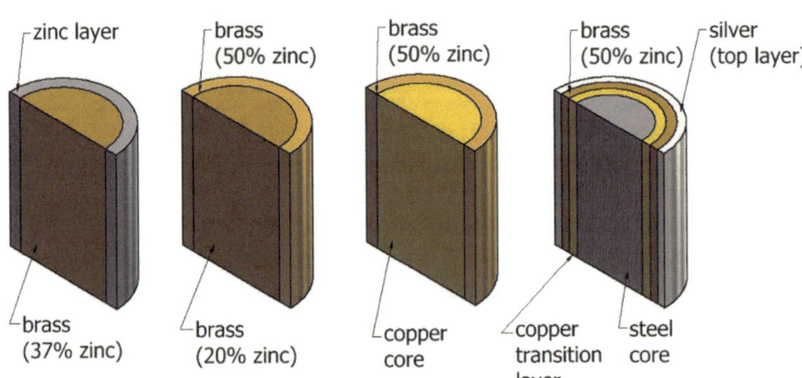

Figure 9.
Selected combinations of multi-component tool electrodes used for WEDM in the case of increased quality requirements for the machined surface.

Type of specific application	Recommended wire electrode composition	The advantage of practical application
Power machining (high MRR)	• Steel wire coated Ms and Cu • Cu wire with Ms coating • Galvanized brass wire	• Better rinsability • A higher wire electrode feed rate
Very small workpiece thickness	• Steel wire coated with Ms and Cu • Graphite coated wire	• Higher resistance to breaking the wire • Better rinsability
Carbide machining and hardly machinable alloys	• Steel wire coated with Ms and Cu • Cu wire with Ms coating	• Higher resistance to breaking the wire • Higher output energy
Machining under different angles so-called conical machining	• Steel wire coated with Ms and Cu • Ms alloy	• Higher resistance to breaking the wire • Increased wire elasticity

Table 2.
Basic properties of composite multi-component tool electrodes and their practical application in specific cases of WEDM.

Their significant disadvantage, compared to single-component compact tool electrodes, is again too high a price. **Table 2** provides an overview of the properties of composite multi-component tool electrodes, including their practical application for specific purposes.

Based on this review, it is evident that composite multi-component tool electrodes provide a number of advantages over conventional single-component compact electrodes. The decisive advantage, however, is their higher tensile strength, which allows the application of higher tension forces *Fw*. In this way, the amplitude of the vibration of the tool electrode can be substantially reduced, thereby achieving a significantly higher quality of the machined surface in terms of its geometric accuracy. All this can be achieved while maintaining the acceptable productivity of the electroerosion process [28]. But the problem is their high price. Therefore, from the point of view of economic efficiency for WEDM in practice, the standard compact single-component tool electrodes continue to be used. However, their limiting factor is the relatively low tensile strength. Therefore, no further significant improvements can be expected in this respect while maintaining an acceptable price of the applied material. It is therefore necessary to draw attention to other possibilities of increasing the geometric accuracy of the machined surface after WEDM. One of the acceptable options is to apply an innovative intelligent control system for generated electrical pulses during the electroerosion process.

6. Analysis of current approaches in the construction of electrical pulse generators used for WEDM

In the past, dependent generators were often used as a source of impulses In the past, dependent generators were often used as a source of impulses. Their running consists in repeated recharging and discharging the capacitor. With this control system discharges the capacitor is normally powered from a DC voltage source, which is connected in parallel to the circuit. Discharging the capacitor occurs when the voltage reaches a breakover value. The size of the breakover voltage depends mainly on the contamination of the dielectric and on the electrode distance. Subsequently, the control system instructs the servo drive to maintain the required working gap size based on the evaluation of the voltage conditions at the discharge location. Changing the time ratios within each discharge also changes their frequency and total discharge energy. From this comes the term "dependent pulse generators."

Thus, it is clear from the above principle that these types of pulse generators allow very short discharges to be produced, while the discharge duration t_i is 10^{-4}–10^{-7}. These are relatively simple construction equipment. For these types of pulse generators, it is required to connect the workpiece as an anode and a tool electrode as a cathode. This type of connection is used because of the need for less material loss from the tool electrode during the electroerosion process. By using a DC power source in a given circuit, the ions are only moved in one direction. This provides a suitable precondition for the formation of corrosion of the eroded particles, which is considered an undesirable phenomenon. In addition, a significant disadvantage of the above-mentioned types of electrical pulse generators is the limited control of the shape and frequency of the discharges, low machining productivity, but also a relatively high wear of the tool electrode [29]. Therefore, these types of electrical pulse generators are no longer used in modern electroerosive equipment.

New types of electrical pulse generators are constantly being developed to continually improve production quality and productivity. These allow, for example, a variable selection of individual electrical discharge parameters, regardless of the actual ratios in the working gap. In addition, the new types of electrical pulse generators have a much longer discharge time than the dependent generators, while lowering the operating voltage. Some types even allow changing the polarity of the discharges during the electroerosion process. In these types of electrical pulse generators, ion conductivity predominates, with the workpiece being normally engaged as a cathode and a tool such as an anode. They are also referred to as independent generators because they allow variable variations in the electrical discharge pulse amplitudes, their polarity, frequency, and so on, regardless of the current situation in the working gap. In technical practice, there are several types of independent generators. For example, a rotary generator. It represents a dynamo that is powered by an asynchronous motor. An essential part of this type of generator is also a semiconductor diode. Its task is to prevent the breakove in the opposite direction. In this type of generator, the tool electrode in the circuit engages as an anode and the workpiece as a cathode. **Figure 10** shows a schematic diagram of an independent electrical pulse generator of a rotating member electroerosive equipment including its volt-ampere characteristic.

These types of independent generators allow relatively high performance, *i.j.* high values of material removal rate (MRR up to 5000 $mm^3.min^{-1}$) at constant frequency of electrical discharges. Therefore, in practice, they are mainly used for roughing operations. However, for the finishing operations, an additional RL generator is required, which is essentially their main disadvantage.

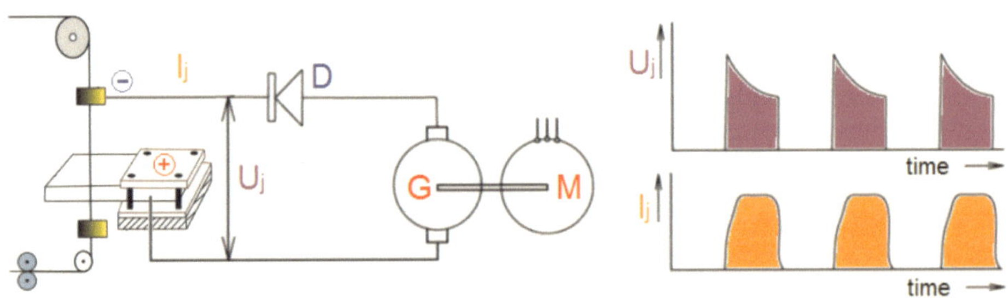

Figure 10.
Scheme of an independent rotary member electrical pulse generator and its volt-ampere characteristic.

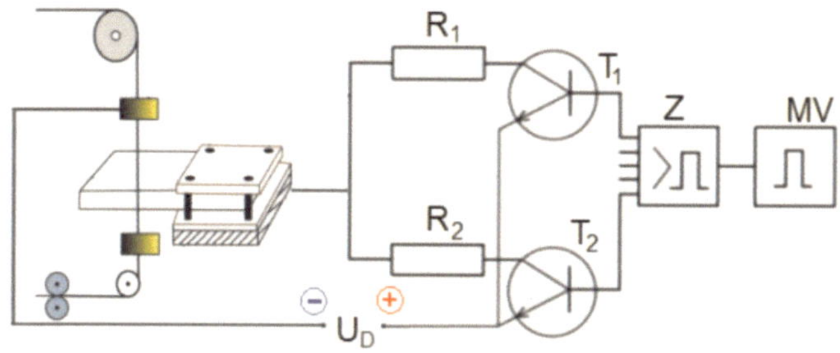

Figure 11.
Schematic of a semiconductor independent electric pulse generator used in modern electroerosion machines.

Higher levels are represented by semiconductor generators. Thanks to the use of semiconductor elements, their main advantage is high reliability, but in particular the possibility to change the parameters of electric discharge in a wide range of values. They allow changing frequencies in the range of 50–500 Hz. Their basic structural element is the semiconductor pulse generator, the so-called multivibrator (MV), which supplies pulses to the amplifier *Z*. This then drives the amplified pulses power transistors T1, T2. Their number determines the amount of current required to be delivered to the discharge location. The frequency of the electrical pulses and their power parameters is determined by the multivibrator. **Figure 11** shows a schematic of an independent generator of electrical impulses electroerosive equipment with semiconductor devices.

Microcomputer controlled generators are currently the most widely used type of independent electrical pulse generator used in state of the art electroerosion machines. These independent semiconductor generators are considered second generation generators. Their main advantage is the application of alternating electric voltage to the discharge, resulting in a reduction of the working gap and the associated reduction in the volume of material withdrawn. They allow a wide range of electrical discharge parameters to be set, with the frequency range of the electrical discharges varying from 0.5 to 50 kHz. The main advantage of the practical application of this type of independent electric pulse generator is the demonstrably less heat affected zone of the eroded area. At the same time, its application can significantly reduce the extent of corrosive effects occurring during the electroerosion process. **Figure 12** shows a diagram of an electroerosion equipment electrical pulse generator that is controlled by a microcomputer.

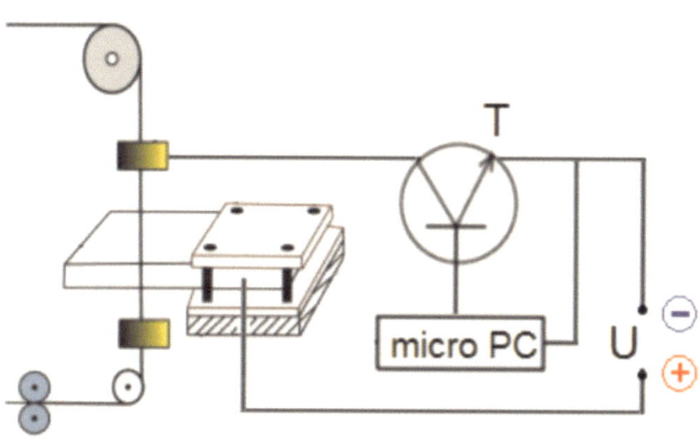

Figure 12.
A diagram of an electrical pulse generator controlled by a microcomputer.

7. Adaptive control system of generated electrical pulses for WEDM

In practice, there are several cases where all of the above possibilities have been used to increase the geometric accuracy of the machined surface in terms of the application of the specific properties of the wire tool electrodes for WEDM. However, despite the application of modern control systems of generated electrical pulses, not all of the expected requirements for the achieved surface finish in terms of geometric accuracy are always met. In this case, one option is to modify the control system of generated electrical pulses. However, it should be pointed out that this is a substantial intervention in the traditionally used system of generated electrical impulses during the electroerosion process.

Another of the requirements for WEDM is, in addition to achieving high quality machined surface in terms of geometric accuracy, also increasing the performance of the electroerosion process. These goals can only be met with the help of highly sophisticated online monitoring systems. At present, information that is derived from the actual value of the electrical discharge parameters is used to control the electroerosion process. In particular, the size of the voltage, current and working gap are monitored. However, setting the current value of these parameters does not take into account all the phenomena that occur in the working gap. They result in the formation of wire electrode vibrations [30]. However, the direct measurement of the amplitude size of the vibration of the wire tool electrode in real life conditions of the electroerosive machine is a problem. One solution is to measure it through one of the indirect methods. Subsequent inclusion of a given parameter as one of the monitored parameters during the WEDM in the form of an input parameter into the process control of generated electric pulses will allow for a new dimension in the field of increasing the productivity of the electroerosion process and the achieved surface quality. At the same time, by monitoring the parameter, a substantial increase in the level of intelligent adaptive control WEDM can be achieved [31]. With its help, it is also possible to detect and then by appropriately adjusting the generated electrical pulses to eliminate the occurrence of unwanted electrical discharges that increase the amplitude of the wire electrode vibration. These informations are very valuable because it allows the use the strategy of an adaptive electrical discharge control to eliminate the adverse impact of inappropriate electrical discharge parameter settings. Improper setting of electrical discharge parameters results in a loss of electroerosion process stability, a decrease in productivity for WEDM, but also a deterioration in the quality of the machined surface. Since a stable electroerosion process is characterized by a constant and uniform vibration of the wire tool electrode with very low amplitude, it is necessary that this deviation at the point of contact of the electrode with the workpiece is regarded as one of the regulatory parameters that ensures the stability of the electroerosion process.

As mentioned above, due to the generation of electrical discharges with inappropriate parameter settings, wire tool electrode vibrations occur. During the electroerosion process, the thin wire electrode primarily vibrates in the X and Y directions. The magnitude of its vibration amplitude is directly subordinated to the frequency and intensity of the generated electrical pulses. **Figure 13** shows the amplitude and direction of vibration of the wire tool electrode during the electroerosion process.

The individual parameters of the generated electrical pulses are currently set with respect to achieving the highest possible efficiency and productivity of the electroerosion process. However, these parameters do not take into account the vibration of the wire tool electrode, which has a significant contribution to the geometric inaccuracy of the machined surface.

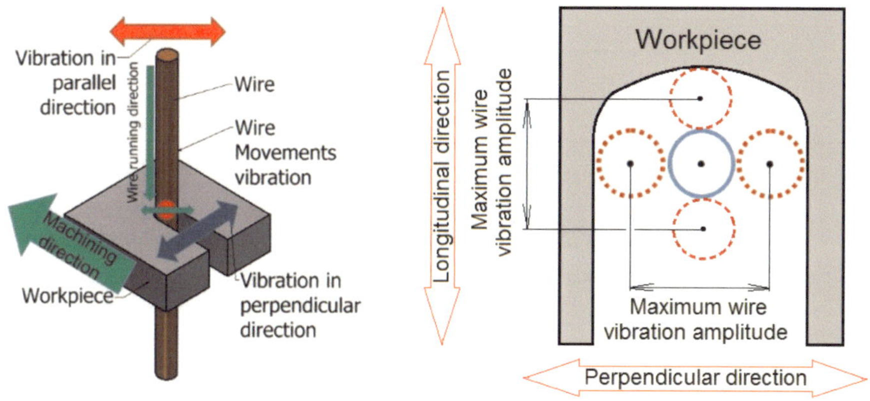

Figure 13.
Amplitude and direction of vibration of the wire tool electrode during WEDM.

As mentioned above, the magnitude of the vibration amplitude of the wire tool electrode is dependent on the size of the wire tensioning force and the current electrical discharge parameters. Of these, priority is given to the frequency of generated electrical discharges. However, based on the results of several investigations, it has been shown that the vibration of the wire electrode in the *X* axis direction becomes slightly higher than the *Y* axis vibration [32]. However, in terms of consequences, the vibration of the wire electrode in the *X* axis direction has a significantly lower effect on the geometric inaccuracy of the machined surface because they are generated in the feed direction. However, the problem is the vibrations that are generated transversely to the wire electrode feed, *i. j.* in the *Y* axis direction. Based on the experimental investigations carried out, it has also been shown that the magnitude of the vibration amplitude of the tool electrode is not directly proportional to the discharge frequency. As can be seen from the graph in **Figure 14**, its maximum value is reached when applying the critical frequency of generated electrical discharges.

However, the critical frequency of the generated electrical discharges, in addition to the electrical discharge parameters, also depends on other parameters, such as the thickness of the material being machined, the diameter of the wire electrode,

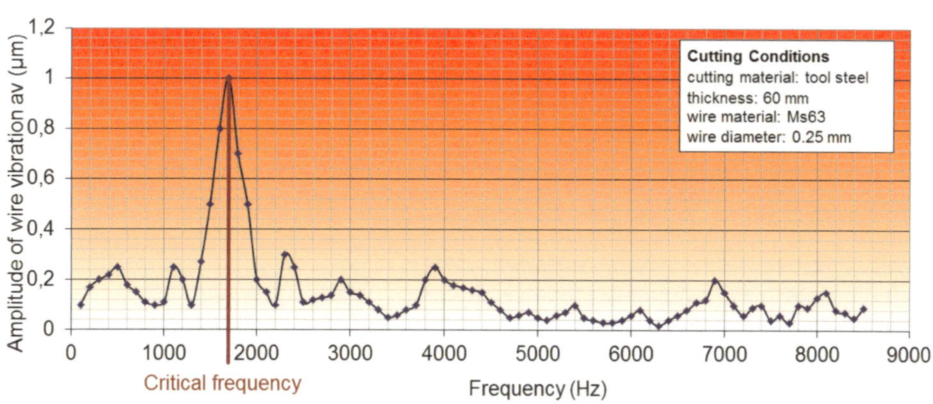

Figure 14.
Dependence of vibration amplitude of wire tool electrode on frequency of generated electrical discharges during WEDM.

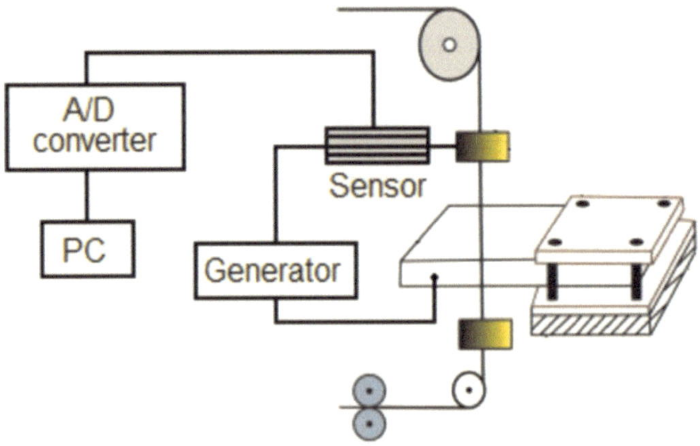

Figure 15.
The principle of indirect measurement of the vibration amplitude of a wire tool electrode during the electroerosion process.

its tension force, and many other parameters. Therefore, the critical frequency of generated electrical discharges during WEDM cannot be implicitly determined.

Since the critical frequency of vibration of the wire tool electrode during the electroerosion process cannot be implicitly determined, the only way to identify it is to monitor the magnitude of the vibration amplitude of the tool electrode It is possible to apply a number of methods to continuously measure the magnitude of the vibration amplitude of a wire electrode during the electroerosion process. Each of these methods has a number of advantages, but also disadvantages. One suitable indirect method for measuring the vibration amplitude of a wire electrode during an electroerosion process that is also applicable to electroerosion machines is the method of acoustic emission signals. Its value can be accurately determined in practice using suitable sensors (**Figure 15**).

An indirect measurement of the vibration amplitude of the tool electrode during WEDM requires a separate approach, since a thin wire is used as a tool in this machining method. The decisive factor in the indirect measurement of the vibration amplitude of the tool electrode is the appropriate positioning of the sensors. Barot et al. [33] research has been conducted in this area who presented a contactless indirect method of measuring the amplitude of the tool electrode vibration by acoustic emission (AE). This is based on comparing the relative intensities of the electromagnetic discharge signals measured by the Hall sensors. However, this method has its limitations because of the need for a magnetic field concentrator to compensate for exponential signal attenuation. At the same time, it requires complicated processing of high frequency electromagnetic signals, which is additionally to be obtained at speeds of up to 30 MHz. Another type of indirect measurement of vibration amplitude of the tool electrode was applied by Okada et al. [34]. The high accuracy of the measurement of the given parameter was achieved by distributing the discharge energy in the wire electrode during WEDM together with direct high-speed digital monitoring of the working gap size. However, this measurement method is only applicable in laboratory conditions. Its application in real electroerosion machine conditions during production is very complicated and therefore impractical.

A special concept of indirect measurement of the vibration amplitude of the tool electrode was performed by Kozochkin et al. [35], and Mahardika et al. [36]. Measurement of the vibration amplitude of the tool electrode was performed by induced discharge with respect to the velocity of the acoustic wave propagated in the machined material (**Figure 16**).

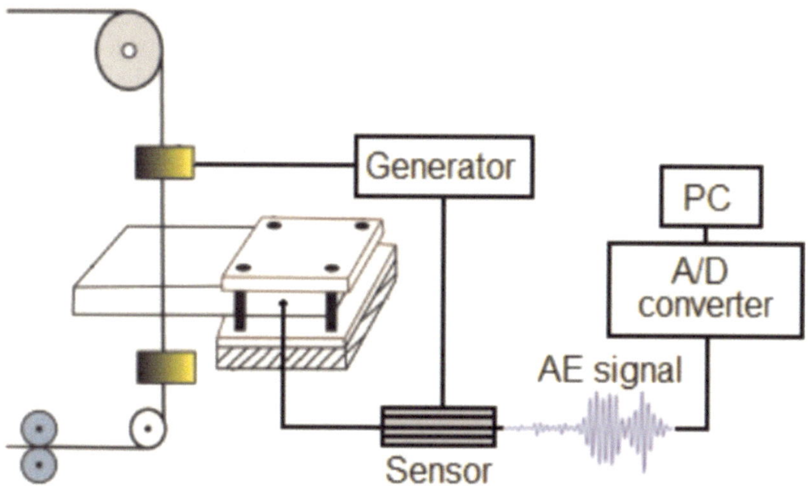

Figure 16.
Measuring the vibration amplitude of the tool electrode by means of an acoustic wave propagating in the machined material.

Smith and Koshy [37] in this indirect method of measuring the amplitude of the tool electrode vibration, they used sensors with a resonance frequency of 20 MHz. However, the measurements performed have shown that this method is only suitable for individual isolated electrical discharges. However, for cyclically repeated discharges, the acoustic waves overlap each other. This results in an unreliable estimate of time delays, which is again impractical for real operation under electroerosion machine conditions.

Another suitable method for indirectly measuring the vibration amplitude of a tool electrode during WEDM appears to be a method of measuring acoustic emission propagating in a tool wire electrode. **Figure 17** demonstrates the appropriate location of the sensors to measure the AE propagated in the tool during WEDM.

The sensors AE may be disposed at one of the ends of the wire tool electrode near the guide rollers or on both at the same time. Since there are cases during

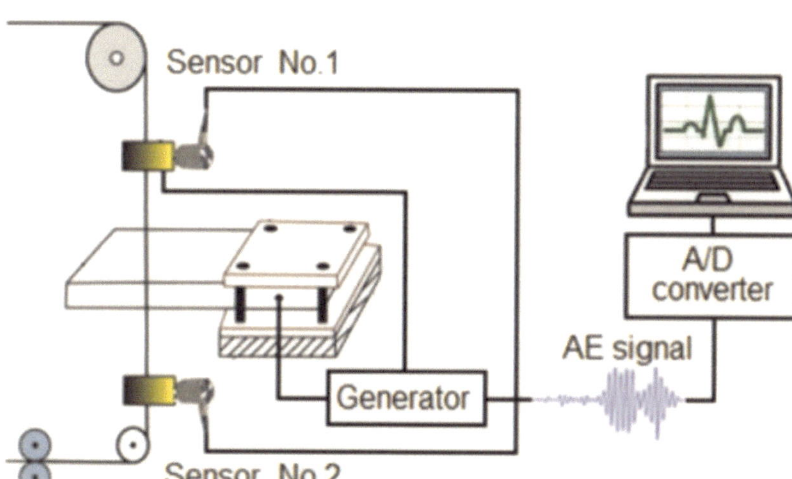

Figure 17.
Measuring the amplitude of the vibration of the tool wire electrode through an acoustic wave propagating in the tool during WEDM.

electrical discharge machining, where the amplitude of the vibration of the wire electrode at one of its ends is slightly increased due to the high thickness of the material being machined or the specific values of the electric discharge parameter settings, it is preferable to install sensors at both the top and bottom of the lead electrode. In a case of only one sensor is applied either at the top or bottom of the wire lead, we could observe distorted values. In this indirect measurement method, electromagnetic interference (EMI) overlap may occur in some cases with the AE signals being sensed. However, this is not a disturbing element in this case, since in both cases it is essentially a noise.

As the decisive criterion for the validity of the recorded data is the location of the sensors, it is necessary to consider the alternative of the AE combination of sensors for the complexity of the solution. In addition, if some research points to some of the advantages of locating the sensors on the wire electrode guide, other on the machined material. In **Figure 18**, a combined way of positioning sensors for AE measurement can be seen. Sensor no. 1 is located on an electroerosion machine in the region of the upper guide of the tool electrode. Sensor no. 2 is located on the workpiece.

However, based on the results of several researches, it was shown that sensor no. 1 placed in the upper guide electrode guide area, indicated more reliable results, as sensor no. 2 placed on the workpiece. At the same time, it is preferable to install AE sensors in the area of the wire tool electrode in terms of practical application. If the AE sensor is located on the workpiece, it must always be re-installed after each workpiece positioning.

However, based on the results of several researches, it was shown that sensor no. 1 placed in the upper conduction of electrode area, indicated more reliable results, as sensor no. 2 placed on the workpiece. These systems allow relatively effective determination of the optimal parameters of the electroerosion process with respect to the required quality of the machined surface. The design of the adaptive control system is implemented based on the principle of self-organization using methods and elements of artificial intelligence. **Figure 19** shows a schematic diagram of the connection of AE sensors to an adaptive electroerosion machine that will eliminate unwanted tool electrode vibrations during WEDM.

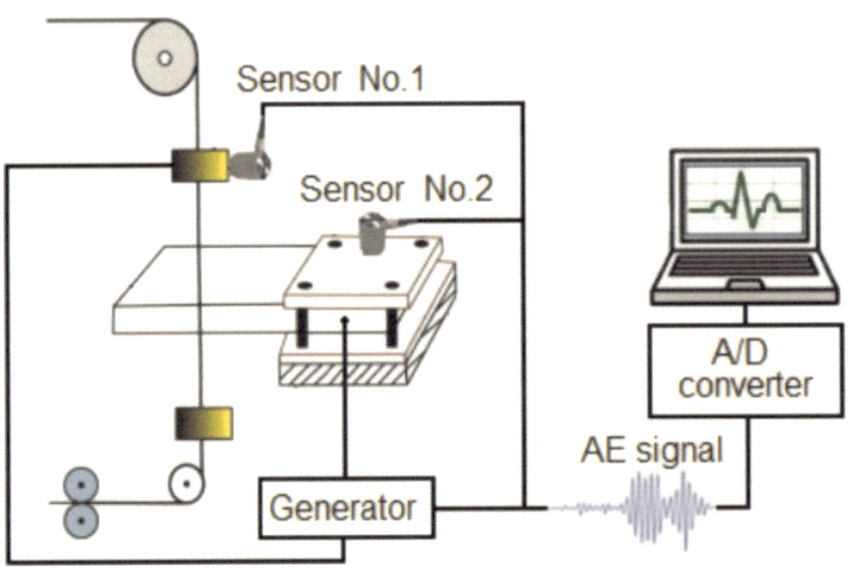

Figure 18.
Method of combined positioning of sensors for measuring AE during WEDM.

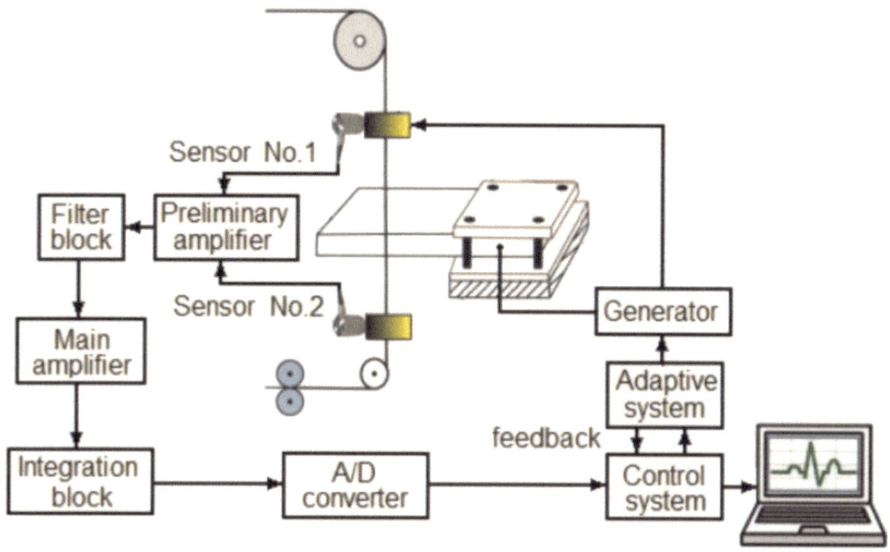

Figure 19.
Principal block diagram of the proposed adaptive system to eliminate unwanted vibrations of the wire tool electrode during WEDM.

The signals received from the AE sensors will be transmitted to the ACD converter. Subsequently, the modified information will be imported into the control system of the electroerosion machine. Based on this input information, it adjusts the electrical discharge parameters by increasing or decreasing their frequency and intensity to minimize unwanted tool electrode vibration.

At the same time, the characteristic feature of the adaptive system for controlling the frequency and intensity of electric discharge during WEDM is the possibility of using optimization techniques based on process algorithms. It is appropriate to apply algorithms that guarantee high convergence in the process of identifying the optimum. To ensure the ideal functionality of the control system of generated electrical pulses, there is a need for a mechanism to be implemented in the system to enable the desired selection of the optimization criterion. This means that in real operation would be possible to choose a priority optimization criteria focused on achieving high quality machined surface, high productivity electroerosion process, high efficiency electroerosion process, eventually their combination. To do this, an expert system based on a large information database is needed. Its suitable connection with the CNC control system of the electroerosion machine would enable efficient operation not only in serial but also piece production.

8. Conclusion

The aim of the book chapter "Intelligent control system of generated electrical pulses at discharge machining" is to provide a comprehensive set of knowledge in the field of intelligent control of generated electrical impulses for WEDM. As is generally known, generated electrical impulses with inappropriate parameters have a negative impact not only on the quality of the machined surface but also on the overall efficiency of the electroerosion process. In addition, since many input parameters change during WEDM, the electroerosion process may become unstable at any time. However, by integrating state-of-the-art monitoring and adaptive control technologies in the field of electroerosion process, not only process stability and performance, but also the quality of the machined surface can be substantially increased. The application of an intelligent control system for generated electrical pulses during WEDM based on electrical discharge input information can effectively prevent the occurrence of an unstable condition or stop the electroerosion process. Although a large number of input factors enter the electroerosion process, the implementation of an intelligent control system for generated electrical pulses is possible through electronic signals. In the case of electroerosion equipment normally produced, the electric discharges generated are controlled on the basis of actual conditions in the working gap. The control system of a traditional CNC electroerosion machine adjusts the performance characteristics of the pulse generator by measuring the average values of electrical voltage and current in the working gap according to predetermined reference values. However, in order to meet the demanding criteria imposed on the quality of the machined surface in terms of achieved geometric accuracy, monitoring of only the mentioned parameters is insufficient. Therefore, the book chapter highlights the importance of monitoring in addition to the established process characteristics such as voltage and current, or the size of the working gap, and the importance of monitoring other process characteristics. Due to the existence of deficiencies reflecting the lack of geometric accuracy of the machined surface, a phenomenon has been identified that causes the poor quality. It is a tool electrode vibration. Although modern electroerosion machines are equipped with algorithms that can to some extent eliminate this unwanted phenomenon, but not at a level that completely eliminates it. Since it has been shown, based on the results of several studies, that the maximum variation in flatness of the machined surface is largely due to the maximum amplitude of vibration of the tool electrode, it is necessary to look for ways to eliminate it. Based on the results of experimental research, it has also been demonstrated that the maximum vibration amplitude of the wire tool electrode is achieved with a specific combination of several factors. However, these cannot be precisely determined. The only solution for identifying its size is to apply one of the measurement methods. The book chapter describes in detail the indirect method of measuring the amplitude of the tool electrode vibration through AE. At the same time it describes possible ways of installing sensors, as well as structure of interconnection of individual components of proposed system. A characteristic feature of the proposed intelligent control system performance parameters of electric discharge during WEDM is its flexibility and openness to the real conditions of practice. Based on an extensive database of information, as well as a rapid and precise exchange of information with an external environment, the system will enable the electroerosion process to be managed with respect to the optimum operation of the electroerosive device according to the individually selected optimization criteria.

Acknowledgements

The authors would like to thank the grant agency for supporting research work the projects VEGA 1/0205/19.

Conflict of interest

The authors declare no conflicts of interests.

Author details

Ľuboslav Straka* and Gabriel Dittrich
Department of Automotive and Manufacturing Technologies, Faculty of Manufacturing Technologies of the Technical University of Kosice with a seat in Prešov, Presov, Slovakia

*Address all correspondence to: luboslav.straka@tuke.sk

References

[1] Qudeiri JEA, Saleh A, Ziout A, Mourad AHI, Abidi MH, Elkaseer A. Advanced electric discharge machining of stainless steels: Assessment of the state of the art, gaps and future prospect. Materials. 2019;**12**:907

[2] Yan MT, Lin TC. Development of a pulse generator for rough cutting of oil-based micro wire-EDM. ISEM XVIII. Procedia CIRP. 2016;**42**:709-714

[3] Świercz DO, Świercz R. EDM—Analyses of current and voltage waveforms. Mechanik. 2017;**2**:1-3

[4] Barik SK, Rao PS. Design of pulse circuit of EDM diesinker. International Research Journal of Engineering and Technology. 2016;**3**(5):2762-2765

[5] Scott D, Boyina S, Rajurkar KP. Analysis and optimization of parameter combination in wire electrical discharge machining. International Journal of Production Research. 1991;**29**:2189-2207

[6] Su JC, Kao JY, Tarng YS. Optimization of the electrical discharge machining process using a GA-based neural network. Journal of Advanced Manufacturing Technology. 2004;**24**: 81-90

[7] Tarng YS, Ma SC, Chung LK. Determination of optimal cutting parameters in wire-eletrical discharge machining. International Journal of Machine Tools and Manufacture. 1995; **35**:1435-1443

[8] Sarkar S, Sekh M, Mitra S, Bhattacharyya B. Modeling and optimization of wire electrical discharge machining of γ-TiAl in trim cutting operation. Journal of Material Processing Technology. 2007;**205**: 376-387

[9] Puri AB, Bhattacharyya B. An analysis and optimization of the geometrical inaccuracy due to wire lag phenomenon in WEDM. International Journal of Machine Tools & Manufacture. 2003;**43**:151-159

[10] Liao YS, Yub YP. Study of specific discharge energy in WEDM and its application. International Journal of Machine Tools & Manufacture. 2006; **44**:1373-1380

[11] Mahapatra S, Patnaik A. Parametric optimization of wire electrical discharge machining (WEDM) process using Taguchi method. Journal of the Brazilian Society of Mechanical Sciences and Engineering. 2006;**28**:422-429

[12] Jin Y, Kesheng W, Tao Y, Minglun F. Reliable multi-objective optimization of high-speed WEDM process based on Gaussian process regression. International Journal of Machine Tools & Manufacture. 2008;**48**:47-60

[13] Yan BH, Tsai HC, Huang FY, Long L. Chorng. Examination of wire electrical discharge machining of Al2O3p/6061Al composites. International Journal of Machine Tools & Manufacture. 2005;**45**:251-259

[14] Yan MT, Lai YP. Surface quality improvement of wire-EDM using a fine-finish power supply. International Journal of Machine Tools & Manufacture. 2007;**47**:1686-1694

[15] Kopac J. High precision machining on high speed machines. Journal of Achievements in Materials and Manufacturing Engineering. 2007; **24**(1):405-412

[16] Shahruz SM. Vibration of wires used in electro-discharge machining. Journal of Sound and Vibration. 2003;**266**

[17] Altpeter F, Roberto P. Relevant topics in wire electrical discharge

machining control. Journal of Materials Processing Technology. 2004;**149**: 147-151

[18] Gokler MI, Ozanozgu AM. Experimental investigation of effects of cutting parameters on surface roughness in the WEDM process. International Journal of Machine Tools & Manufacture. 2000;**40**:1831-1848

[19] Hewidy MS, El-Taweel TA, El-Safty MF. Modelling the machining parameters of wire electrical discharge machining of Inconel 601 using RSM. Journal of Materials Processing Technology. 2005;**169**:328-336

[20] Hašová S, Straka L'. Design and verification of software for simulation of selected quality indicators of machined surface after WEDM. Academic Journal of Manufacturing Engineering. 2016;**14**(2):13-20

[21] Mičietová A, Neslušan M, Čilliková M. Influence of surface geometry and structure after non-conventional methods of parting on the following milling operations. Manufacturing Technology. 2013;**13**:199-204

[22] Ferdinandov N et al. Increasing the heat-resistance of X210Cr12 steel by surface melting with arc discharge in vacuum. In: Metal 2018, 27th International Conference on Metallurgy and Materials, Brno. 2018. pp. 1097-1102

[23] Straka L', Čorný I, Piteľ J. Prediction of the geometrical accuracy of the machined surface of the tool steel EN X30WCrV9-3 after electrical discharge machining with CuZn37 wire electrode. Metals. 2017;**7**(11):1-19

[24] Panda A et al. Considering the strength aspects of the material selection for the production of plastic components using the FDM method. MM Science Journal. 2018;**2018**(12): 2669-2672

[25] Straka L', Čorný I, Piteľ J, Hašová S. Statistical approach to optimize the process parameters of HAZ of tool steel EN X32CrMoV12-28 after die-sinking EDM with SF-Cu electrode. Metals. 2017;**7**(2):1-22

[26] Zhang W, Wang X. Simulation of the inventory cost for rotable spare with fleet size impact. Academic Journal of Manufacturing Engineering. 2017;**15**(4): 124-132

[27] Straka L', Hašová S. Prediction of the heat-affected zone of tool steel EN X37CrMoV5-1 after die-sinking electrical discharge machining. Proceedings of the Institution of Mechanical Engineers Part B: Journal of Engineering Manufacture. 2016;**9**:1-12

[28] Swiercz R et al. Optimization of machining parameters of electrical discharge machining tool steel 1.2713. In: AIP Conference Proceedings, EM 2018, 13th International Conference Electromachining 2018, Bydgoszcz. 2018. Article no. 020032

[29] Salcedo AT, Arbizu PI, Perez CJL. Analytical modelling of energy density and optimization of the EDM machining parameters of inconel 600. Metals. 2017; 7(5):166

[30] Wang X. An experimental study of the effect of ultrasonic vibration assisted wire sawing on surface roughness of SiC single crystal. Academic Journal of Manufacturing Engineering. 2017;**15**(4): 6-12

[31] Melnik YA et al. On adaptive control for electrical discharge machining using vibroacoustic emission. Technologies. 2018;**6**(4):96

[32] Habib S, Okada A. Experimental investigation on wire vibration during fine wire electrical discharge machining process. International Journal of Advanced Manufacturing Technology. 2016;**84**:2265-2276

[33] Barot RS, Desai KP, Raval HK. Experimental investigations and monitoring electrical discharge machining of Incoloy 800. Journal of Manufacturing Engineering. 2017;**12**(4): 196-202

[34] Okada A, Uno Y, Nakazawa M, Yamauchi T. Evaluations of spark distribution and wire vibration in wire EDM by high-speed observation. CIRP Annals—Manufacturing Technology. 2010;**59**:231-234

[35] Kozochkin MP, Grigor'ev SN, Okun'kova AA, Porvatov AN. Monitoring of electric discharge machining by means of acoustic emission. Russian Engineering Research. 2016;**36**(3):244-248

[36] Mahardika M, Mitsui K, Taha Z. Acoustic emission signals in the micro-EDM of PCD. Advanced Materials Research. 2008;**33-37**:1181-1186

[37] Smith C, Koshy P. Applications of acoustic mapping in electrical discharge machining. CIRP Annals—Manufacturing Technology. 2013;**62**: 171-174

Chapter 2

Conceptual Design Evaluation of Mechatronic Systems

Eleftherios Katrantzis, Vassilis C. Moulianitis and Kanstantsin Miatliuk

Abstract

The definition of the conceptual design phase has been expressed in many different phrasings, but all of them lead to the same conclusion. The conceptual design phase is of the highest importance during the design process, due to the fact that many crucial decisions concerning the progress of the design need to be taken with very little to none information and knowledge about the design object. This implies to very high uncertainty about the effects that these decisions will have later on. During the conceptual design of a mechatronic system, the system to be designed is modeled, and several solutions (alternatives) to the design problem are generated and evaluated so that the most fitting one to the design specifications and requirements is chosen. The purpose of this chapter is to mention some of the most widely used methods of system modeling, mainly through hierarchical representations of their subsystems, and also to present a method for the generation and evaluation of the design alternatives.

Keywords: conceptual design, mechatronic design, hierarchical modeling, concept evaluation, mechatronic abilities, Choquet integral, criterion interactions

1. Introduction

The current mechatronic systems acquire very advanced capabilities based on the evolution of the mechatronics enabling technologies and the mechatronic design methodology. The enhanced intelligence of the mechatronic systems and the increased complexity are identified; however, these changes drive to completely new characteristics and capabilities of mechatronic systems supporting the new generation of production systems, e.g., these devices evolved from the simple monitoring to self-optimizing their performance. On top of that, mechatronics enhanced the application domains from manufacturing to biomechatronics and micromechatronics.

The development of mechatronic products and systems requires concurrent, multidisciplinary, and integrated design approaches. This chapter deals with methods and models used during the mechatronic design and more specifically with the design evaluation. A method for concept generation and evaluation is presented. The criteria used as well as the mathematical foundations of the method will be presented and analyzed. More specifically, the mechatronic criteria based on the mechatronic abilities as well as their scoring will be described using a systematic approach. Aggregation of the new criteria will be performed using a nonlinear fuzzy integral. The use of the mechatronic design index in a chosen application task will be presented.

The next section of this chapter is a two-part state of the art. First, the conceptual design of mechatronic systems and different techniques and approaches for the system modeling and representation during this phase are discussed. Later, the state of the art concerning concept evaluation methodologies and indexes is presented. A method for concept evaluation is proposed in Section 3 of the chapter. In the final section, an exemplary case study of the conceptual design phase of a mechatronic object is presented and discussed in the Conclusions section.

2. State of the art

2.1 System modeling and hierarchical representations

Conceptual model creation of a mechatronic object to be designed is the actual task for industrial production systems that usually operate with modern CAD/CAM systems [1–4]. According to [1], in the object life cycle, the conceptual design is made just before the phase of creating the detailed design, when the object's concrete mathematical model is created and numeric calculations are realized.

Nowadays, there are many definitions of conceptual design and corresponding methods and models that are used at the conceptual design phase analyzed in [5]. An opinion given by Hudspeth [5] is that conceptual design is more about what a product might be or do and how it might meet the expectations of the manufacturer and the customer. M. J. French defines conceptual design as the phase of the design process when the statement of the problem and generation of a broad solution to it in the form of schemes is performed [6].

The US Department of Transportation's FLH Project Development and Design Manual [7] states that conceptual studies (CS) are typically initiated as needed to support the design planning and programming process. The CS phase identifies, defines, and considers sufficient courses of action (i.e., engineering concepts) to address the design needs and deficiencies initially identified during the planning process. This phase advances a project proposed in the program to a point when it is sufficiently described, defined, and scoped to enable the preliminary design and technical engineering activities to begin. The CS studies and preliminary design phases are performed in conjunction and concurrently with the environmental process, which evaluates environmental impacts of the engineering proposals resulting from the conceptual studies and preliminary design phases.

Functional modeling technology was researched and applied to represent concept design knowledge by [8, 9] presented a function-behavior-structure (FBS) ontology representation process for concept design in different domains and emphasized the reasoning mechanism with the FBS ontology for knowledge representation.

Borgo et al. [10] proposed an ontological characterization of artifact behavior and function to capture the informal meanings of these concepts in the engineering practice and characterize them as part of a foundational ontology. The function-cell-behavior-structure (FCBS) model to better comprehend representation and reuse of design knowledge in conceptual design was proposed by Gu [11]. A hierarchical two-layer concept is given here, i.e., two knowledge-representing layers—the principle layer and the physical layer—are presented in the FCBS model. The principle layer is utilized here to represent the principle knowledge. Case modeling is employed in the physical layer to integrate the structural information and behavioral performances of the existing devices that apply the design principles represented by the functional knowledge cells (FKCs).

A formal definition of the concept design and a conceptual model linking concepts related to design projects are proposed by Ralph and Wand [12]. Their definition of design incorporates seven elements: agent, object, environment, goals, primitives, requirements, and constraints. The design project conceptual model is based, here, on the view that projects are temporal trajectories of work systems that include human agents who work to design systems for stakeholders and use resources and tools to accomplish this task. Ralph and Wand [12] demonstrate how these two conceptualizations can be useful by showing that (1) the definition of design can be used to classify design knowledge and (2) the conceptual model can be used to classify design approaches.

An approach for using hierarchical models in the design of mechatronic systems is presented by Hehenberger in [13]. To master the mechatronic design approach, a hierarchical design process is proposed. The models cover the different views on a system as well as the different degrees of detailing. The utilization and proper combination of solution principles from different domains of mechatronics allow an extended variety and quality of principal solutions, where hierarchical models serve as very important tools for complex design tasks. Analysis of different mechatronic design concepts is also conducted in the work. The approach is demonstrated by studying the activities during the design process of synchronous machines.

Another approach used in mechatronic design is knowledge-based engineering (KBE) described by Sobieszczanski-Sobieski in [14]. The main structures for extended KBE application are design process and design models. The models contain specific aspects such as product structure as a whole and its fragments, engineering calculations, and analysis with ability of integration with external systems, design requirements, and decision-making processes. This object-oriented approach makes it possible to speed up the process of generating the source code of design models from the extended KBE and supports multidisciplinary design optimization. Knowledge-based hybrid intelligent systems, namely, imperialist competitive algorithm, artificial neural networks, genetic algorithms, and particle swarm optimization, are also used in tunnel design and construction processes and described in [15].

Mathematical models used in design and modeling of mechanical structures of mechatronic systems are described in [16]. The examples of car suspension system modeling were presented in this work. The approach of interactive design and production evaluating of a manufacturing cell is described in [17].

The use of hierarchical system (HS) formal construction and HS coordination technology in conceptual design of mechatronic objects is proposed and described by Miatliuk [18].

2.2 Concept evaluation and generation

The mechatronic design quotient (MDQ) [19–21] was proposed as a multicriterion measure for assisting decision-making in mechatronic design. In MDQ seven criteria were incorporated: meeting task requirements, reliability, intelligence, matching, control friendliness, efficiency, and cost. These criteria are aggregated by means of the Choquet Integral—a nonlinear fuzzy integral that can be used for assisting decision-making with interactive criteria [22]. Guidelines for the concept evaluation using these criteria were presented in [21], where four alternatives of an industrial fish cutting machine were evaluated using a hierarchical classification of the aforementioned criteria and the Choquet integral as the aggregator.

The mechatronic multicriteria profile (MMP) [23] includes five main criteria (machine intelligence quotient, reliability, complexity, flexibility, and cost of manufacture and production) for the mechatronic concept evaluation. The MMP criteria are defined in such a way that the assessment of the alternatives with respect to each criterion results from measurable sizes and does not depend entirely on the designer's judgment and experience. In [23], the proposed method is applied to the conceptual design of a visual servoing system for a 6-DOF robotic manipulator, and the Choquet nonlinear fuzzy integral is used for the aggregation of the criteria. Three different aggregation techniques, the Choquet integral, the Sugeno integral, and a fuzzy-based neural network, were tested and compared for the design evaluation of a quadrotor mechatronic system [24].

The mechatronic index vector (MIV) introduced in [2] consists of three criteria, intelligence, flexibility, and complexity. The attributes of every criterion are analyzed and formulated. The intelligence level of a system is determined by its control functions, and the structure for information processing of mechatronic systems is used to model intelligence. A technique to measure the flexibility of manufacturing systems was used for the estimation of the flexibility of a mechatronic product. The various types of flexibility were classified in three main categories, namely, product flexibility, operation flexibility, and capacity flexibility. The complexity was modeled using seven elements. Various models for aggregating the criteria were proposed and compared including t-norms, averaging operators [2], and the discrete Choquet integral [25].

Ferreira [26] proposed a decision support tool based on a neural network to provide suggestions for early design decisions based on previous solutions. In the same manner, the mechatronic design indicator (MDI) was proposed [27] as a performance indicator based on a neuronal network of radial basis functions.

In [28], Moulianitis proposed a new mechatronic index for the evaluation of alternatives. The proposed criteria that make up the mechatronic index were mainly extracted from the collective knowledge presented in the multi annual roadmap (MAR) for robotics in Europe [29] and adapted by considering the recent advancements in mechatronics. The discrete Choquet integral is used for the aggregation of the evaluation scores, while also taking into account the correlations between criteria. The criteria and the aggregation method of the mechatronic index are presented in detail in the following sections.

In recent years, some research has been focused in the automated generation of system architecture concepts for mechatronic design problems. In [30], an automated generation and evaluation method for feasible and ranked physical architectures is proposed. First, the components that can realize specified system functions are identified and combined with the use of a unified knowledge model and dynamic programming methods. Then, the criteria are realized, and the system architectures are evaluated based on the technique for order preference by similarity to an ideal solution (TOPSIS).

An integrated principle solution synthesis method which achieves the automated synthesis of multidisciplinary principle solutions but also solves the undesired physical conflicts among synthesized solutions was proposed in [31].

In [32], a model-based research approach for an integrated conceptual design evaluation of mechatronic systems using SysML software is proposed and applied to the design of a two-wheel differential drive robot to find the optimal combination of component alternatives for specific evaluation goals.

3. Concept evaluation in mechatronic design

In this section, the necessary steps for concept evaluation are presented and described. The process is described in terms of a flowchart in **Figure 1**.

The definition of the design specifications and requirements is a perquisite for a more complete evaluation. The designer can also capitalize on a well-defined set of

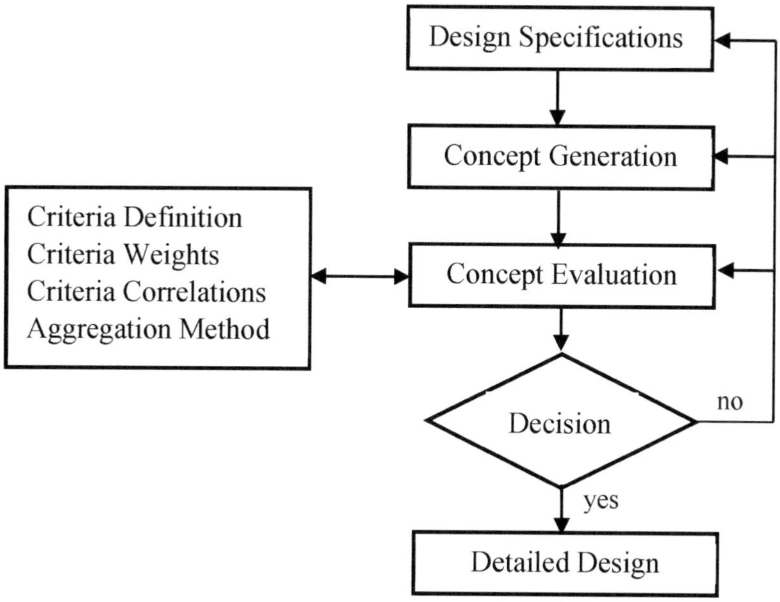

Figure 1.
The proposed evaluation process during the conceptual design phase.

requirements and specifications in order to generate functional concepts that will later be evaluated. In order for the evaluation to take place, four steps are necessary. The criteria that are to be used for the evaluation of the alternatives must be chosen and defined. Then, the weights (importance) of the said criteria must be decided, as well as the interactions between them. The design alternatives are then evaluated with respect to each criterion, and those scores are aggregated using a discrete Choquet integral in order for the final scoring values to be derived. The alternative with the highest score is considered to be the most suitable to the design problem, given the specific requirements and other design environment characteristics (design team experience, knowledge, etc.). At this point, the designer has the choice to select the alternative that will be further developed during the detailed design phase or to review some of the previous step and reevaluate the design alternatives. The steps presented in **Figure 1** are presented in more detail in the following paragraphs. The final section of this chapter is an exemplary case study where the proposed evaluation method is used for the evaluation of a mechatronic system.

3.1 Design specifications

In the early stage of defining the mechatronic system/product to be designed, the engineering specifications and constraints are determined. The quality function deployment (QFD) is a well-known and widely used method for deriving the design specifications and constraints by realizing customers' needs and requirements. In [33], the classification of design specifications to demands and wishes is proposed, and, by extension, the further grading of wishes is based on their importance over the design process. The determination and the recording of design specifications are of great importance to the concept generation and evaluation processes. During the concept generation process, design specifications and constraints can help the designer to easily recognize any unfeasible concepts that will not satisfy the minimum requirements, while during the concept evaluation process, the designer can consult the list of customer requirements and design specifications for the evaluation criterion selection and the determination of their weights.

3.2 Concept generation

Concept generation is the process where possible solutions to the design problem are realized based on design specifications and functions that the system/product must accomplish. In order for a generated concept to be considered a feasible solution to the design problem, the two following conditions must be satisfied: (a) the concept meets at least the minimum design specifications, and (b) it includes the necessary software and hardware components [28]. Concepts can be represented in different ways, such as sketches or flow diagrams, function hierarchies, textual notes, or table representations. However, regardless of which way a concept is represented, enough detail must be developed to model performance so that the functionality of the idea can be ensured [1]. The quality and the thoroughness of the generated alternatives mainly depend on the available information and knowledge about the design problem at that early stage of the process and the experience of the design team.

3.3 Concept evaluation

In the concept evaluation stage, the concepts that have been generated are being evaluated with respect to some criteria. The performance of each alternative with respect to each criterion is rated, and these scores are aggregated with a specified aggregation function. The purpose of the evaluation is to support the decision-making process of the designer and help him/her to choose the best concept for further analysis in the detail design stage. In this section, the basic ideas behind each step of the process will be described, and different ways of implementing them will be mentioned. More emphasis will be placed on describing the foundations and methods (criterion definition, aggregation function, correlations between criteria) to be applied in the exemplary case study that follows in the next section.

3.4 Criterion definition

A criterion can be thought of as a measure of performance for an alternative [34]. As we saw earlier in the state of the art section, more than a few criteria and combinations of those criteria have been proposed for the evaluation of mechatronic concepts. The type and number of criteria to be used are up to the designer and the stakeholders that take part in the design process. The selection of criteria could depend on the designer's experience and personal judgment; the available information and knowledge about the design problem; the alternatives and the abstraction level of their description, design specifications, and customer requirements; and whether the designer wants to incorporate the interactions among the criteria in the decision-making process.

For the purposes of this chapter, the mechatronic abilities proposed by Moulianitis in [28] as criteria for the evaluation of mechatronic systems will be used in the exemplary case study. The criteria are based on the collective knowledge presented in the Multi Annual Roadmap (MAR) for Robotics in Europe [29]. The mechatronic abilities found in [28] are the following:

- Adaptability
- Configurability

- Decisional autonomy
- Dependability
- Interaction ability
- Motion ability
- Perception ability

Abilities provide a basis for setting performance metrics and for application providers to specify desired levels of system performance [29]. These ability levels are adapted to mechatronic criteria, and a scoring scale for concept evaluation with respect to these criteria is presented [28]. Different scaling types for scoring and evaluation of criteria have been proposed, and the dispute concerning the superiority among them has been discussed in [35]. The way the criteria are mapped to scoring methods affects the evaluation results, meaning that a suitable mapping could lead to more realistic results. However, this work is outside the scope of this chapter and is left for future work.

For the scoring of alternatives, it is assumed that the progression of the levels advances the characteristics of the system linearly, so a linear interpolation is used to map each level to a score. The criteria are scaled in the same universe of discourse, with the lowest possible value being equal to zero and the highest possible equal to one. The scores of the intermediate levels are assigned linearly between zero and one. In the following, the scaling of the criteria according to [28], as well as short descriptions of the criteria, is provided.

3.4.1 Adaptability

Fricke [36] defined adaptability as the ability of a system to adapt in order to deliver intended function ability under varying conditions by changing the values of the design parameters either actively (online) or passively (off-line). In [29], adaptability is defined as the ability of the system to adapt itself to different work scenarios and different environments and conditions. Adaptability is often mixed with configurability and decisional autonomy but is differentiated by configurability in the sense that adaptation is mostly devoted to the parameter change rather than to the structure change. The difference between autonomous decision and adaptation is that adaptation takes place over time based on an accumulation of experience, while decisional autonomy is a result of environmental perception by means of sensors and cognitive mechanisms.

Adaptability can be broken down to five ability levels, starting from Level 0 when the system has no ability to adapt and reaching up to Level 4 when the process of adaptation is carried out by multiple agents. In the three intermediate levels, the system behavior is self-evaluated, and the need for parameter adaptation is recognized (Level 1), and in addition, individual parameters can be altered based on local performance assessment (Level 3), and in Level 4 the adaptation concerns multiple parameter changes. The levels of adaptability and the scaling for the scoring of each level are presented in **Table 1**.

3.4.2 Configurability

Configurability is the ability of a system to alter its configuration to perform different tasks. As it is stated in [29], configurability must be carefully distinguished

Adaptability level	Normalized score
0	0
1	0.25
2	0.5
3	0.75
4	1

Table 1.
Levels and scaling for adaptability.

from adaptability and decisional autonomy which relate to how a robot system alters its responses (adaptability) and how it changes its behavior as it performs an operating cycle. At the highest level (Level 4) of configurability, the system is able to sense changes in its environment's conditions that are not pre-programmed and alter its configuration in response to those changes. At the lowest levels, the system has a single and non-alterable configuration (Level 0), and at Level 1 the user is responsible for the definition of the system configuration at the beginning of each cycle of operations. As we go up to Level 3, the system can alter its configuration autonomously from a predetermined set of alternative built-in configurations, and in Level 4 the system is able to alter its configuration in response to changing conditions that are not pre-programmed or predetermined. The score scaling for each configurability level is shown in **Table 2**.

3.4.3 Decisional autonomy

A feature of many mechatronic systems is the devolution of functional responsibility to the system, freeing the operator or user to pay attention on the higher-level functions associated with the deployment and applicability of the system [37]. In order to enhance the decisional autonomy of a system, it should be equipped with heuristics, machine learning capabilities, logic tools, etc. The same as before, the leveling starts from Level 0 for systems with no ability to take decisions, and it goes up as the system enhances its ability to take decisions. At first, the system is fully dependent on user decisions (Level 1), the system makes decisions to choose its behavior from a predefined set of alternatives based on basic sensing and user inputs (Level 2), and at Level 3 the system is able to process the inputs from the user and the sensing unit and makes decisions continuously, while in Level 4 moment-to-moment decisions about the environment are taken. Level 5 introduces an internal model of the environment to the system in order to support the system's decision-making process, and when in Level 6, the sequence of predefined subtasks is decided in a way, so it accomplishes a higher-level task. Level 7 means that the system can adapt its behavior to accommodate task constraints, and Level 8

Configurability level	Normalized score
0	0
1	0.25
2	0.5
3	0.75
4	1

Table 2.
Levels and scaling for configurability.

translates to the alteration of the strategy as the system gathers new information and knowledge about the environment. In the two higher levels, decisions about actions are altered within the time frame of dynamic events that occur inside the environment (Level 9), and in Level 10, system compensation in real-time events is enabled by the alteration of the tasks themselves. The levels and the corresponding scores are presented in **Table 3**.

3.4.4 Dependability

Dependability of mechatronic units is defined as the qualitative and quantitative assessment of the degree of performance, reliability, and safety taking into consideration all relevant influencing factors [38]. The higher the level of dependability of a system, the more reliable this system is. Seven levels of dependability are defined as follows. At Level 0, there is no system ability to predict any failures. At Level 1, dependability is measured only by estimations of the mean time between failures, and the system has no real ability to detect or prevent those failures. At Level 2, the system has the ability to diagnose a failure and enter safe mode operation, while at Level 3 the system is able to diagnose a number of failures and recover from a proportion of them. If the system has the added ability to predict the consequences to its tasks caused by the diagnosed failures, then it has reached Level 4. At Level 5, the system can communicate its failures to other systems in order to rearrange the aggregate sequence of tasks and keep its mission dependable, and as we reach Level 6, the system is able to predict a failure and act to prevent it. Dependability levels and scores are presented in **Table 4**.

3.4.5 Interaction ability

It is the ability of a system to interact physically, cognitively, and socially either with users, operators, or other systems around it [29]. In the concept of human adaptive mechatronics (HAM) [39], the goal is to design a mechatronic system that includes the user in the control loop and modifies the functions and the structure of user-machine interface to improve the human's operational skills. Interactivity is considered in most of the modern mechatronic systems to facilitate either the

Decisional autonomy level	Normalized score
0	0
1	0.1
2	0.2
3	0.3
4	0.4
5	0.5
6	0.6
7	0.7
8	0.8
9	0.9
10	1

Table 3.
Levels and scaling for decisional autonomy.

Dependability level	Normalized score
0	0
1	0.17
2	0.33
3	0.5
4	0.67
5	0.83
6	1

Table 4.
Levels and scaling for dependability.

operation or the maintenance and repair. Six levels were used for the modeling of interaction ability. Level 0 entails that no ability for interaction exists. The lowest level where human-system interactions are present is Level 1, where the operation of the system can be interrupted at any time by the user. When human-machine interaction is possible even if the user and the system are isolated, we are at Level 2, and if the system's workspace is divided into safe and unsafe zone for human interaction, then Level 3 of interaction ability is reached. At Level 4, human-system synergy is considered, while the system checks for dangerous motions or forces that could be harmful to the human. At the highest level, that is, Level 5, recognition of the conditions under which the system should have a safe mode behavior based on detection of uncertainty is enabled. Interaction ability levels and their scoring are presented in **Table 5**.

3.4.6 Motion ability

In [28], motion ability is considered to categorize the different types of motion control. Open-source 3D printers are systems with Level 1 motion ability, while robotic vacuum cleaners are presenting abilities up to Level 5. Motion ability levels and the scores are presented in **Table 6**. At the lowest level (Level 0), the system presents no motion ability. As we go up to Level 1, the system accomplishes predefined motions in a sequence using open-loop control, while the use of closed-loop control for motion in a predefined manner is considered to be at Level 2 of motion ability. Level 3 offers the ability for constrained position or force options integrated to the motion control, while a reactive motion describes Level 4. Optimization of a set of parameters and planning of its motion based on said optimization translate to Level 5 of motion ability.

Interaction ability level	Normalized score
0	0
1	0.2
2	0.4
3	0.6
4	0.8
5	1

Table 5.
Levels and normalized values for interaction ability [10].

Motion ability level	Normalized score
0	0
1	0.17
2	0.4
3	0.6
4	0.8
5	1

Table 6.
Levels and scores for motion ability [10].

3.4.7 Perception ability

In order for mechatronic systems to be capable of operation in unstructured, dynamic environments, multiple sensors and methods for sensor fusion and environment recognition are integrated into them [40]. The perception ability of a system is associated with its ability to understand and sense its working environment. The leveling of the perception ability extends to eight levels in total. The same with all the previous mechatronic abilities, Level 0 means that there is no ability to perceive data. If critical data are collected using sensors and the behavior of the system is directly altered, then the system is at Level 1 of perception ability. If the collected data are first processed and the behavior of the system is then indirectly altered, we are at Level 2. At Level 3, multiple sensors are being used to create a unified model of the surrounding environment, while at Level 4 system is able to extract features of the environment by sensing only a region of it. Being a Level 5 system goes with the ability to process the sensing data in order to extract information features that help with better environment interpretation. At Level 6 objects are identified using an object model, and at the highest level, Level 7, processed data are used in order to infer about properties of the environment. Perception ability levels and their scoring are given in **Table 7**.

3.4.8 Criteria weights

Criteria weights express the designer's preference for the importance of each criterion in the assessment of the alternatives. Various methods for assigning weights to criteria have been proposed in the relevant literature [1, 33, 34, 41]. The

Perception ability level	Normalized score
0	0
1	0.14
2	0.29
3	0.43
4	0.57
5	0.71
6	0.86
7	1

Table 7.
Levels and normalized scores for perception ability [10].

most commonly used weight assignment technique for concept evaluation in the mechatronic design process is the direct rating of each criterion weight by the decision-makers (direct rating, point allocation, numerical scale) [1, 41, 42]. The eigenvector method, proposed by Saaty in [43], is a simple method that uses pairwise comparisons and ratings between criteria in order to formulate the weight of each individual criterion. In [34, 41], the reader can find a more in-depth analysis of different weight rating methods and how to choose the most suitable method depending on the decision problem. In the process of assigning values to criteria weights, the design team should make an effort to consider the design specifications and customer requirements and try to reflect them on the assigned values.

3.5 Criteria correlations and aggregation method

Choquet integral is a nonlinear fuzzy integral, which has been proposed and used for the aggregation of interacting criteria [22]. The integral allows for the designer to incorporate interactions into the evaluation process by providing weighting factors (weights) both for the criteria and the correlations between each subset of criteria. Considering the set of criteria $X = \{x_1, x_2, \dots, x_n\}$, the concept of a fuzzy measure [44] is defined.

A fuzzy measure on the set X of criteria is a set function $\mu : P(X) \to [0,1]$ satisfying the following axioms:

i. $\mu(\varnothing) = 0 \textit{ and } \mu(X) = 1$

ii. $A \subset B \subset C \textit{ implies } \mu(\mathrm{A}) \le \mu(\mathrm{B})$.

In this context, $\mu(\mathrm{A})$ represents the weight of importance of criterion A [22]. By expressing the weighting factors of each subset of criteria, the interactions between criteria can be taken into account during the aggregation. Four types of interactions between criteria are presented in this chapter. A positive interaction (or correlation) means that a good score in criterion $\mathrm{x_i}$ implies a good score in criterion $\mathrm{x_j}$ and vice versa, while a negative correlation between interacting criteria means that a good score in criterion $\mathrm{x_i}$ implies a bad score in criterion $\mathrm{x_j}$ and vice versa. If a criterion has a veto effect on the evaluation process, a bad score in criterion $\mathrm{x_i}$ results in a bad global score. A pass effect, on the other hand, implies that a good score in criterion $\mathrm{x_i}$ results in a good global score. In **Table 8**, the four types of interactions between criteria and the relations between their weighting factors are presented.

Given the set of criteria X and the fuzzy measure μ, the Choquet integral of a function $f : X \to [0,1]$ with respect to μ is defined by [23]

$$C_\mu = C_\mu(f(x_1), \dots f(x_n)) := \sum_{i=1}^{n} \left[\left(f\left(x_{(i)}\right) - f\left(x_{(i-1)}\right) \right) \mu\left(\mathrm{A}_{(i)}\right) \right] \tag{1}$$

Interaction	Relation
Positive correlation	$\mu(x_i, x_j) < \mu(x_i) + \mu(x_j)$
Negative correlation	$\mu(x_i, x_j) > \mu(x_i) + \mu(x_j)$
Veto effect	$\mu(T) \approx 0 \textit{ if } T \subset X - \{x_i\}$
Pass effect	$\mu(T) \approx 1 \textit{ if } T \subset X, x_i \in T$

Table 8.
Correlations between criteria.

where $(.)_{(i)}$ indicates that the indices have been permuted so that $0 \leq f(x_1) \leq \ldots \leq f(x_n) \leq 1, f(x_{(0)}) = 0$, and $A_{(i)} = [x_{(i)}, \ldots, x_{(n)}]$.

or equally [45]

$$C_\mu = \sum_{i=1}^{n} f(x_{(i)}) [\mu(\mathrm{A}_{(i)}) - \mu(\mathrm{A}_{(i+1)})] \tag{2}$$

where $(.)_{(i)}$ indicates that the indices have been permuted so that the criteria values are sorted in ascending order, such that $0 \leq f(x_1) \leq \ldots \leq f(x_n) \leq 1$, $A_{(i)} = [x_{(i)}, \ldots, x_{(n)}]$, and $A_{(n+1)} = 0$.

Marichal [45] also proposed an axiomatic characterization of the integral to motivate its use in applications. The expression he ended up with is the Choquet integral in terms of the Mobius representation:

$$C_\mu = \sum_{i \in N} a(i) \bullet f(x_{(i)}) + \sum_{\{i,j\} \subseteq N} a(i,j) \bullet [f(x_{(i)}) \wedge f(x_{(j)})] + \ldots \tag{3}$$

where $a(i) = \mu(i)$ and $a(i,j) = \mu(i,j) - [\mu(i) + \mu(j)]$.

4. Case study: automated BMP system

The proposed method is used for the concept evaluation of an automated biomethane potential (BMP) measurement system. Before the presentation of the evaluation of the system, it should be mentioned that all the scoring values of alternatives, the criteria weights, and the criteria correlation weights were given intuitively by the authors of this chapter. This is not ideal, since the design team should strive for a more analytic and objective approach to the evaluation process. However, the study of the cognitive mechanisms that take place during the evaluation and the consideration of different ways of scoring criteria and criteria weights is left for future work. The purpose of this chapter is to focus more on the general methodology for concept evaluation and not so on its specifics.

Production of biogas from different organic materials is an interesting source of renewable energy. The biomethane potential (BMP) of these materials has to be determined to get insight in design parameters for anaerobic digesters [46]. The test is conducted by placing an active inoculum and a sample of a substrate in a sealed container vessel and measuring the amount of the gas produced [47]. The basic steps followed in a BMP test are:

1. The substrate and the inoculum are placed in the sealed reactor vessel.

2. Stirring (not continuous) and heating (the temperature is held constant in the range 20–90°C, depending on the test mixture) of the test mixture to enhance biogas production. The production of biogas starts at this step and continuous until the end of the experiment.

3. Absorption of CO_2. The biogas is transferred to a reactor vessel containing (liquid) NaOH. This allows for the dissolution of the CO_2 in the NaOH solution, and the remaining gas volume is representative of the CH_4 present in the biogas.

4. Gas flow measurement. The CH_4 is transferred to the flow measurement system. The volume of gas produced in a specific time interval is quantified, usually under the fluid displacement principle.

5. Data analysis and reporting.

An average BMP test can last for more than 30 days, during which the test system must run without disruptions. The BMP test can be done manually, where human interference in 24-h intervals is needed in order for the gas flow measurement to take place. The whole process can also be automated, where no human interaction is needed for the gas flow measurement, and the analysis and representation of the results take place in real time during the experimental process. Two commercial products that automate the BMP test are the AMPTS II by Bioprocess Control Systems [48] and the Biogas Batch Fermentation System by Ritter [49]. The following scheme represents the basic functions that need to be accomplished for the automation of the process. The functions presented in the scheme in **Figure 2** are incorporated in both products (AMPTS II and Biogas Batch Fermentation System).

In **Figures 3** and **4**, the two products are presented in relation to the scheme given in **Figure 2**. They accomplish the same functions but with slightly different components and working principles.

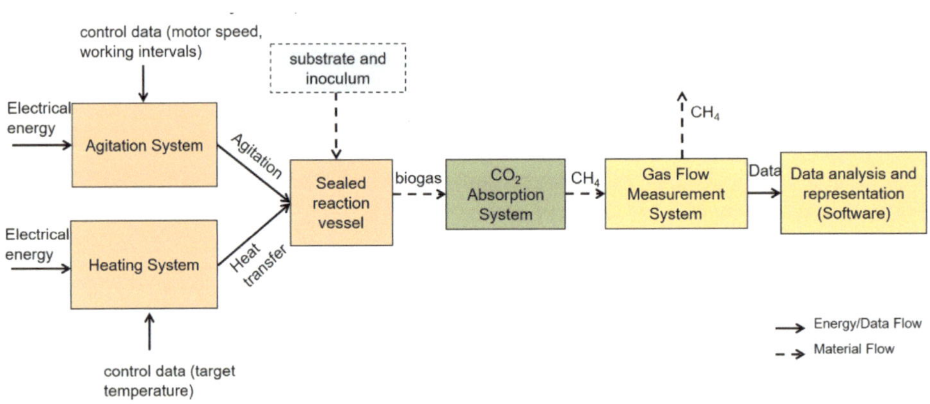

Figure 2.
Subfunctions and flows of material and energy of an automated BMP system.

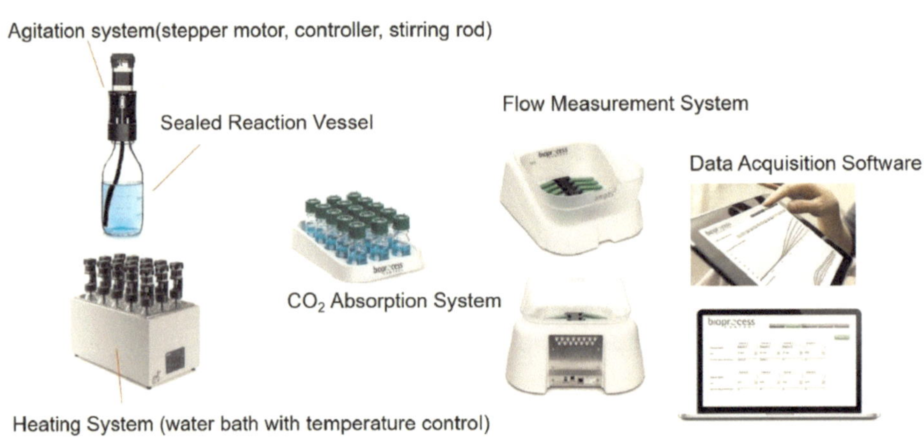

Figure 3.
The AMPTS II system by bioprocess control [48].

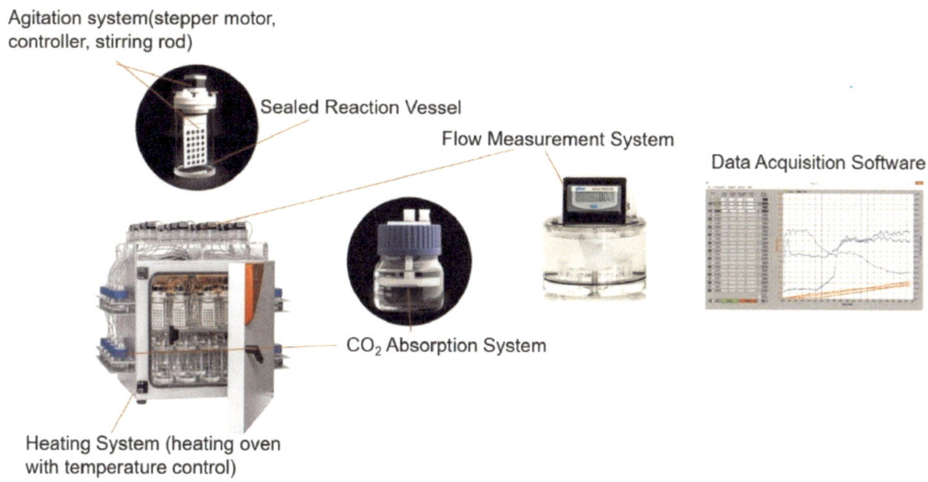

Figure 4.
The biogas batch fermentation system by Ritter [49].

The agitation in the case of the AMPTS is accomplished with an overhead stirrer with a special airtight cap that allows for the stirring of the mixture with a stirring rod coupled to the actuator. Ritter used an overhead stirrer, but the coupling between the actuator and the stirring rod is accomplished with a magnetic coupling. Ritter utilized a heating oven for the temperature control of the mixture, while in AMPTS II a water bath is used. For the ultralow gas flow measurement (ULGFM), Bioprocess Control uses their patented gas flow cell meter which operates based on the liquid displacement and buoyancy working principles, as well as the Hall effect. Ritter's gas flowmeter is based on the same operating principles but utilizes the tipping bucket effect in combination with a Hall effect sensor. For the normalization of the measurement results, a temperature and a barometric pressure sensor are being used. The user can control the speed and the movement direction of the actuator motors through the software.

The main requirements considered by the design team are that the system can reliably operate for the whole duration of the experiment (more than 30 days), the temperature of the mixture inside the reactor vessel is held constant at a prespecified temperature, it offers an automated and accurate flow measurement method, its components and subsystems are gastight in order for the gas to be able to travel through the system without any leakages, and finally the results are presented to the user via a software UI.

After the listing of the design requirements, the design team proceeds with the generation of alternative concepts. For the facilitation of concept generation, the system could be modeled in various ways, such as the scheme shown in **Figure 5**. In **Table 9**, the information about the mechatronic system is presented in terms of energy, material, and information flows [1].

By studying existing automated BMP systems, the design team attempted to model the basic functions of any automated BMP measuring system as a flowchart. This flowchart representation of the system functions is meant to be representative of any automated BMP measurement system and help the design team with the better understanding of the system to be designed.

In [28], a design tree with the subfunctions of a mechatronic object, more specifically an educational firefighting robot, is presented. A representation of the automated BMP system in a manner similar to that design tree is presented in **Figure 5**.

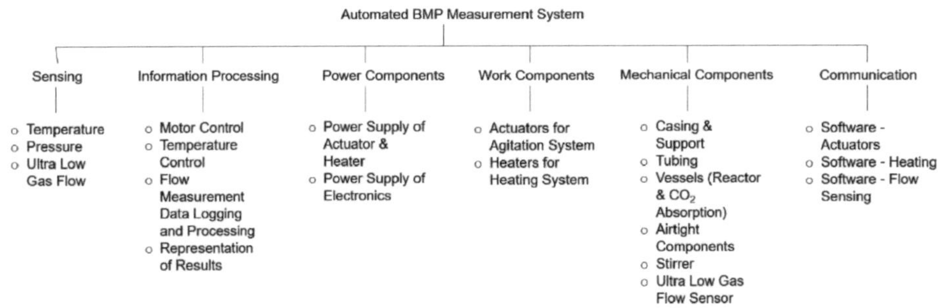

Figure 5.
Design tree for automated BMP measurement system.

Energy flow	Material flow	Information flow
Transformation of electrical power to kinetic energy (agitation)	Flow of gas mixtures through the system	Temperature, pressure, and flow sensing
Transformation of electrical power to thermal energy	Agitation of mixture in the reactor vessels	Control of actuators
		Control of heating element
		Flow sensor data processing and result representation

Table 9.
Energy, material, and information flow representation of the automated BMP system.

The component and component categories that are considered in the design tree in **Figure 5** are the following:

- Sensing: For temperature control inside the reactor vessel, pressure recording and gas measurement.

- Information processing: Software that is responsible for the actuator and temperature control, the data logging and processing of flow measurement values, and the calculation and representation of the experimental results.

- Power components: The power supply of the actuators, the heating elements, and the control and software electronics.

- Work components: The components that produce some kind of work. Mechanical work by the agitation actuators and thermal work by the heating elements.

- Mechanical components: For the casing and structural support of the system, the tubing connections, the vessels (reactor vessel, absorption unit), the stirrer, the airtight components, and the mechanical parts contained within the flow measurement sensor.

- Ultralow gas flow measurement: Working principles incorporated within the gas flow measurement system. In reality, the design of an ultralow gas flow measurement system could be considered by itself as a distinct mechatronic design problem.

- Communication: For the inter process communications between the software and the rest subsystems (actuators, heater, sensors).

Based on the classification of components considered in the design tree of the system (**Figure 5**) and the subsystems considered in the flowchart representation of the system (**Figure 2**), the design team came up with a first set of alternative concepts for the automated BMP system. The alternative components considered are presented in **Table 10**.

As it can be seen in **Table 10**, not all subsystems and components were taken into account in the generation of possible solutions, but this does not mean that the basic requirements of the system are not met by the alternatives that have been chosen. As mentioned in previous sections, the quality and completeness of the solutions depend to a large extent on the available information and knowledge about the design object and the design team experience. It should also be borne in mind that the evaluation of the solutions is an iterative process, the purpose of which is to support the design team in the decision-making processes during the conceptual design phase.

Based on the data of **Table 11**, there are $4 \times 3 \times 2^4 = 192$ possible design solutions. The study of the procedures for reducing the number of solutions lies outside the scope of this chapter and is left for future work. However, some of the factors that may play an important role in rejecting some alternatives without requiring a thorough evaluation process are outlined. The factors are as follows.

- The design team is not familiar with relevant technology [28].
- The assembly/communication between some components is unacceptably complicated or even impossible.
- The cost of some alternatives is too high.
- Some alternatives have already been realized by competitors, and they are not considered innovative enough.
- The alternative does not satisfy the basic system/product requirements.
- The time frame for the design process does not allow for the exhaustive evaluation of the alternatives, and a decision must be made quickly.

Agitation system	Heating system	Temperature sensors	Ultralow gas flow measurement (ULGFM)	Communication
1. Actuator: stepper, DC brushed/ brushless, servo 2. Stirrer: overhead coupled stirrer, magnetic stirrer	3. Water bath heater 4. Heating oven 5. Silicon jacket heater	6. Resistance thermometer detector (RTD) 7. Thermocouple	8. Liquid displacement and buoyancy and Hall sensor 9. Liquid displacement and optical sensor	10. Direct, wired between software and subsystems 11. Radio frequency (Wi-Fi) between software and subsystems

Table 10.
Automated BMP measurement system alternatives based on system subfunctions.

Design alternatives	DA_1	DA_2	DA_3
Actuator	DC brushed	DC brushless	Stepper
Stirrer	Overhead with linear coupling	Overhead with linear coupling	Magnetic
Heating	Water bath	Heating oven	Silicon jacket
Temperature sensors	Thermocouple	RTD	RTD
ULGFM	Liquid displacement and optical sensor	Liquid displacement and buoyancy and Hall sensor	Liquid displacement and optical sensor
Communication	Wired connections	Wired connections	Radio frequencies (Wi-Fi)

Table 11.
Three design alternatives selected for evaluation by the design team.

In the same manner, if an alternative presents really low cost and complexity and the design team is very familiar with the relevant technology, it could be chosen for further development along with the alternatives that will be chosen after the evaluation process. The design team came up with the three design alternatives DA_k $(k = 1, 2, 3)$ presented in **Table 11**.

Each design alternative uses a different type of actuator for the stirring of the mixture. In DA_1 and DA_2, a brushed DC motor and a brushless DC motor, respectively, are used. A stepper motor with the additional components for the open-loop control of the system is being utilized in DA_3. As for the other agitation components, DA_1 and DA_2 are using an overhead stirrer which is coupled with the motor with a linear coupling, e.g., helix coupler, while the third alternative makes use of a magnetic coupling between the actuator and the stirrer. The three design alternatives realize the heating of the mixture inside the reaction vessels in three different ways. The two first alternatives, i.e., DA_1 and DA_2, use a water bath and a heating oven, respectively, for the heating of the mixture, while in DA_3 a silicon rubber heater (etch foil heater) is chosen for the heating purposes. Thermocouples (DA_1) and a RTD sensor (DA_2, DA_3) are chosen as alternative solutions for temperature sensing inside the reactor vessel. The two alternatives that the design team came up with for the realization of the ultralow gas flow measurement system are presented in **Figure 6**.

The first alternative for the ULGFMS, presented in **Figure 6a** and used in design alternatives DA_1 and DA_3, utilizes the liquid displacement discipline and with the use of an optical sensor is able to calculate the flow of the produced gas. As gas

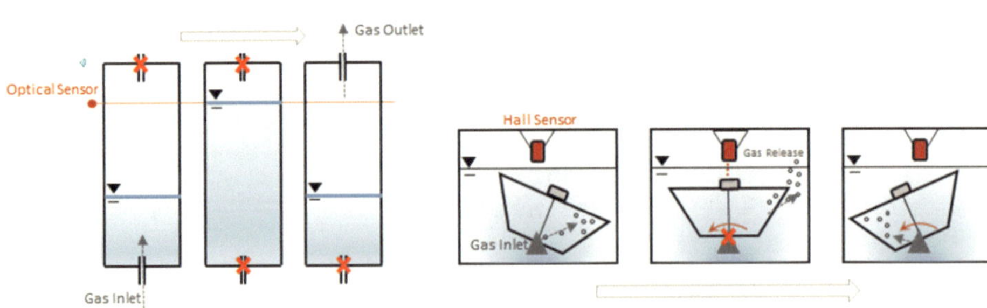

(a) (b)

Figure 6.
Ultralow flow measurement alternatives. (a) The liquid displacement discipline. (b) The tipping bucket mechanism.

enters the chamber, pressure arises, and the liquid level inside the chamber rises until it reaches the point where the optical sensor is pointed at. The sensor records the phenomenon, and the gas is then released from the system, the liquid level drops again, and the process repeats itself. By calibrating the system so that we know the exact gas volume needed for the liquid to reach the optical sensor's level and recording the number of times it reaches that level in a given time interval, we can estimate the gas flow. The second alternative presented in **Figure 6b** and utilized in design alternative DA_2 makes use of a tipping bucket mechanism. A bucket-like chamber is placed inside a container packed with a liquid. Gas bubbles enter the bucket, and when enough of them have gathered, the bucket tips because of the buoyancy. During the tipping motion, a Hall sensor records the phenomenon, and the gas is released. As we can see, the second alternative is very similar to the first one. Finally, for the first two alternatives DA_1 and DA_2, the communication between its subsystems will all be achieved via physical (wired) communication protocols and hardware components, while the third DA_3 alternative is using radio frequencies, namely, Wi-Fi communication between its subsystems.

The next step of the conceptual design phase involves the selection of the appropriate criteria for the evaluation of the design alternatives, the determination of their weights, and the interactions between them. In the example of the automated BMP system, five criteria from the total of seven presented in the previous section are considered for the evaluation, along with the cost and complexity criteria. For the cost criterion, the material resources and man-hours required for the development of the system are estimated. Complexity mainly describes the familiarity of the design team with specific technologies, and complex design process will be based on that fact. It should be noted that higher scores on cost and complexity criteria translate to fewer resources needed for the development of the product and also to less complex designs. The score of each alternative with respect to each criterion is shown in **Table 12**.

The fuzzy measures that represent the weight of each criterion are presented in **Table 13**. Three values were chosen to specify the importance of each criterion $(0.05, 0.105, 0.160)$, with each one corresponding to low, medium, and high importance, respectively. As it is stated in [28], a choice of very low and/or very high values limits the ability to define the correlations between criteria. For example, assuming a low importance equal to $\mu(x_1) = 0.01$ and a high importance equal to $\mu(x_2) = 0.20$, then the positive correlation among them is impossible to be defined since the following constraints must be true: $\mu(x_1) + \mu(x_2) > \mu(x_1, x_2)$ and $\mu(x_1, x_2) > \mu(x_2)$.

The design team wanted to reward systems that display low cost and complexity, and so the two criteria are negatively correlated. The same thinking applies to

Criteria (x_i)	DA_1	DA_2	DA_3
1. Configurability	0.01	0.05	0.3
2. Dependability	0.17	0.17	0.17
3. Interaction ability	0.2	0.2	0.3
4. Motion ability	0.25	0.3	0.17
5. Perception	0.14	0.14	0.14
6. Cost	0.5	0.2	0.4
7. Complexity	0.4	0.3	0.3

Table 12.
Evaluation scores.

Criteria (x_i)	Weight ($\mu(x_i)$)
1. Configurability	$\mu(x_1) = 0.16$
2. Dependability	$\mu(x_2) = 0.05$
3. Interaction ability	$\mu(x_3) = 0.105$
4. Motion ability	$\mu(x_4) = 0.05$
5. Perception	$\mu(x_5) = 0.05$
6. Cost	$\mu(x_6) = 0.16$
7. Complexity	$\mu(x_7) = 0.16$

Table 13.
Criteria weights.

Criteria (x_i, x_j)	Interaction	Set weight ($\mu(x_i, x_j)$)
Cost/complexity	Negative	$\mu(x_6, x_7) = 0.45$
Complexity/configurability	Negative	$\mu(x_1, x_7) = 0.40$
Interaction ability/configurability	Positive	$\mu(x_1, x_3) = 0.20$
Cost/configurability	Negative	$\mu(x_1, x_6) = 0.35$

Table 14.
Criteria correlations and set weights.

Design alternatives	DA_1	DA_2	DA_3
Evaluation scores	0.2995	0.1960	0.3205

Table 15.
Choquet values of the alternative scores.

two other negative correlations between complexity/configurability and cost/configurability. A system with great configurability allows for the user to customize the BMP system easier and faster, which translates to higher interaction ability levels, and thus, the two criteria are positively correlated. All interactions and the corresponding weights are presented in **Table 14**.

The final evaluation scores of each alternative are presented in **Table 15**. The evaluation scores are the Choquet integral values that were calculated based on Eq. (3). Alternative DA_3 scored the highest score, while DA_1 and DA_2 came up second and third, respectively, as it is shown below. DA_1 and DA_3 present similar performance characteristics, with DA_1 having a marginal lead in cost and complexity performance but falls short on configurability performance.

5. Conclusions

In this chapter, some well-known methods for the conceptual analysis and evaluation during the mechatronic design process are discussed, and a method for the evaluation of generated design alternatives is proposed. The proposed design criteria, which were derived from the multi annual roadmap in robotics in Europe, were presented in [28]. Four interactions between criteria are presented and the Choquet integral along with the two additive fuzzy measures used for dealing with the aggregation of the evaluation scores.

The most useful outcomes of this chapter are as follows. (i) The modeling of the system can lead to a better understanding of the problem and make the evaluation process easier and more accurate. (ii) The proposed mechatronic abilities can be utilized in a number of different situations. However, the score scaling of the criteria needs to be further investigated. (iii) The proposed method is there to support the design team on the selection of the most suitable design alternative. The evaluation process and the results obtained from it are dependent on the experience of the design team, the number of people participating in the evaluation, and the available knowledge at the time the decision is made.

Acknowledgements

This research has been co-financed by the European Union and Greek national funds through the Operational Programme for Competitiveness, Entrepreneurship and Innovation, under the call RESEARCH-CREATE-INNOVATE (project code: T1EDK-03148).

Author details

Eleftherios Katrantzis[1*], Vassilis C. Moulianitis[1] and Kanstantsin Miatliuk[2]

1 Department of Product and Systems Design Engineering, University of the Aegean, Ermoupoli, Syros, Greece

2 Department of Automatic Control and Robotics, Bialystok University of Technology, Bialystok, Poland

*Address all correspondence to: lef.katrantzis@syros.aegean.gr

References

[1] Ullman D. The Mechanical Design Process. 1992

[2] Moulianitis VC, Aspragathos NA, Dentsoras AJ. A model for concept evaluation in design—An application to mechatronics design of robot grippers. Mechatronics. 2004;**14**(6):599-622

[3] Shigley JE, Mischke CR. Mechanical Design Engineering Handbook. Elsevier; 2013

[4] Johnson J. Designing with the Mind in Mind Well-Known User Interface Design Rules; 2012

[5] Hudspeth M. Conceptual design in product data management. Des. Eng. Technol. News Deskt. Engineering.

[6] French MJ. Conceptual Design for Engineers. London: Springer London; 2013

[7] FLH. Project Development and Design Manual. US Department of Transportation; 2014 [online]. Available from: https://flh.fhwa.dot.gov/resources/

[8] Bryant CR, Stone RB, McAdams DA. Automated Concept Generation from the Functional Basis of Design. 2001; **573**:1-24. Available from: search.informit.com.au

[9] Gero JS, Kannengiesser U. A function-behavior-structure ontology of processes, in artificial intelligence for engineering design. Analysis and Manufacturing: AIEDAM. 2007;**21**(4):379-391

[10] Borgo S, Carrara M, Garbacz P, Vermaas PE. A formal ontological perspective on the behaviors and functions of technical artifacts. Artificial Intelligence for Engineering Design, Analysis and Manufacturing. 2009; **23**(1):3-21

[11] Gu CC, Hu J, Peng YH, Li S. FCBS model for functional knowledge representation in conceptual design. Journal of Engineering Design. Aug. 2012;**23**(8):577-596

[12] Ralph P, Wand Y. A proposal for a formal definition of the design concept. Lecture Notes in Business Information Processing (LNBIP). 2009; **14**:103-136

[13] Hehenberger P, Poltschak F, Zeman K, Amrhein W. Hierarchical design models in the mechatronic product development process of synchronous machines. Mechatronics. Dec. 2010;**20**(8):864-875

[14] Sobieszczanski-Sobieski J, Morris A, van Tooren MJL. Multidisciplinary Design Optimization Supported by Knowledge Based Engineering. 2015

[15] Koopialipoor M, Fallah A, Armaghani DJ, Azizi A, Mohamad ET. Three hybrid intelligent models in estimating flyrock distance resulting from blasting. Engineering Computations. Jan. 2019;**35**(1):243-256

[16] Azizi A, P. Y.-C.-B. A. of the S. S. Of, and U. Mechanical Structures: Mathematical Modeling, Springer; 2019K

[17] Azizi A, Ghafoorpoor Yazdi P, Hashemipour M. Interactive design of storage unit utilizing virtual reality and ergonomic framework for production optimization in manufacturing industry. International Journal on Interactive Design and Manufacturing. Mar. 2019; **13**(1):373-381

[18] Miatliuk K. Conceptual Design of Mechatronic Systems. 2017

[19] De Silva CW, Behbahani S. A design paradigm for mechatronic systems. Mechatronics. 2013;**23**(8):960-966

[20] Behbahani S, de Silva CW. System-based and concurrent design of a smart mechatronic system using the concept of mechatronic design quotient (MDQ). IEEE/ASME Transactions on Mechatronics. 2008;**13**(1):14-21

[21] Behbahani S, de Silva CW. Mechatronic design quotient as the basis of a new multicriteria mechatronic design methodology. IEEE/ASME Transactions on Mechatronics. Apr. 2007;**12**(2):227-232

[22] Grabisch M. The application of fuzzy integrals in multicriteria decision making. European Journal of Operational Research. 1996;**89**(3): 445-456

[23] Mohebbi A, Achiche S, Baron L. Mechatronic multicriteria profile (MMP) for conceptual design of a robotic visual servoing system. In: Volume 3: Engineering Systems; Heat Transfer and Thermal Engineering; Materials and Tribology; Mechatronics; Robotics, 2014. p. V003T15A015

[24] Mohebbi A, Achiche S, Baron L. Multi-criteria fuzzy decision support for conceptual evaluation in design of mechatronic systems: A quadrotor design case study. Research in Engineering Design. Jul. 2018;**29**(3): 329-349

[25] Moulianitis VC, Aspragathos NA. Design evaluation with mechatronics index using the discrete choquet integral. IFAC Proceedings. Jan. 2010; **39**(16):348-353

[26] Ferreira IML, Gil PJS. Application and performance analysis of neural networks for decision support in conceptual design. Expert Systems with Applications. 2012;**39**(9):7701-7708

[27] Hammadi M, Choley JY, Penas O, Riviere A, Louati J, Haddar M. A new multi-criteria indicator for mechatronic system performance evaluation in preliminary design level. In: 2012 9th France-Japan & 7th Europe-Asia Congress on Mechatronics (MECATRONICS)/13th International Workshop on Research and Education in Mechatronics (REM). 2012. pp. 409-416

[28] Moulianitis VC, Zachiotis G-AD, Aspragathos NA. A new index based on mechatronics abilities for the conceptual design evaluation. Mechatronics. Feb. 2018;**49**:67-76

[29] SPARC. Robotics 2020 Multi-Annual Roadmap MAR ICT-24 ii; 2015

[30] Chen R, Liu Y, Fan H, Zhao J, Ye X. An integrated approach for automated physical architecture generation and multi-criteria evaluation for complex product design. Journal of Engineering Design. Mar. 2019;**30**(2–3):63-101

[31] Qi J, Hu J, Peng Y-H. An integrated principle solution synthesis method in multi-disciplinary mechatronic product conceptual design. Concurrent Engineering. 2018;**26**(4):341-354

[32] Chami M, Bruel JM. Towards an integrated conceptual design evaluation of mechatronic systems: The SysDICE approach. Procedia Computer Science. 2015;**51**(1):650-659

[33] Pahl G, Beitz W. Engineering Design: A Systematic Approach; 2013

[34] Sen P, Yang J-B. Multiple Criteria Decision Support in Engineering Design; 2011

[35] Ishizaka A, Labib A. Review of the main developments in the analytic hierarchy process. Expert Systems with Applications. 2011;**38**(11):14336-14345

[36] Fricke E, Schulz AP. Design for changeability (DfC): Principles to enable changes in systems throughout their entire lifecycle. Systems Engineering. 2005;**8**(4):342-359

[37] Bradley D. Mechatronics and intelligent systems. 2005. pp. 395-400

[38] Kochs H-D, Petersen J. A Framework for Dependability Evaluation of Mechatronic Units; 2004

[39] Yu H. Overview of human adaptive mechatronics. Mathematics and Computers in Business and Economics. 2008:152-157

[40] Luo RC, Chou YC, Chen O. Multisensor fusion and integration: Algorithms, applications, and future research directions. In: Proceedings of the 2007 IEEE International Conference on Mechatronics and Automation, ICMA. Vol. 2007. 2007. pp. 1986-1991

[41] Hwang C-L, Yoon K. Methods for Multiple Attribute Decision Making. Berlin, Heidelberg: Springer; 2012. pp. 58-191

[42] Bottomley PA, Doyle JR, Green RH. Testing the reliability of weight elicitation methods: Direct rating versus point allocation. Journal of Marketing Research. 2003;**37**(4):508-513

[43] Saaty TL. A scaling method for priorities in hierarchical structures. Journal of Mathematical Psychology. 1986;**13**:65-75

[44] Sugeno M. Theory of Fuzzy Integrals and its Application. 1972

[45] Marichal J-L. An axiomatic approach of the discrete Choquet integral as a tool to aggregate interacting criteria. IEEE Transactions on Fuzzy Systems. 2000;**8**(6)

[46] Holliger C et al. Towards a standardization of biomethane potential tests. Water Science and Technology. 2016;**74**(11):2515-2522

[47] Angelidaki I et al. Defining the biomethane potential (BMP) of solid organic wastes and energy crops: A proposed protocol for batch assays. Water Science and Technology. Mar. 2009;**59**(5):927-934

[48] Bioprocess Control. AMPTS II—Methane potential analysis tool [Online]. Available from: https://www.bioprocesscontrol.com/products/ampts-ii/ [Accessed: 01-Jun-2019]

[49] Ritter. Biogas Batch Fermentation System [Online]. Available from: https://www.ritter.de/en/products/biogas-batch-fermentation-system/ [Accessed: 01-Jun-2019]

Chapter 3

Mechatronics for the Design of Inspection Robotic Systems

Pierluigi Rea and Erika Ottaviano

Abstract

Recent trends show how mobile robots are being widely used in security and inspection tasks. This chapter reports the requirements, characteristics, and development of mobile robotic systems for security and inspection tasks to demonstrate the feasibility of mechatronic solutions for inspection of sites of interest. The development of such systems can be exploited as a modular plug-in kit to be installed on a mobile system, with the aim to be used for inspection and monitoring, introducing high efficiency, quality, and repeatability in the addressed sector. The interoperability of sensors with wireless communication constitutes a smart sensor toolkit and a smart sensor network with powerful functions to be used efficiently for inspection purposes. A tele-operated robot will be taken as case of study; it is controlled by mobile phone and equipped with internal and external sensors, which are efficiently managed by the designed mechatronic control scheme.

Keywords: mechatronics, hybrid leg-wheel locomotion, inspection, interoperability, low-cost monitoring

1. Introduction

Robotics applied to inspection and home security is becoming a reality in recent years. Moreover, with the spread diffusion and popularity of robots in everyday life, their use is enormously increased in recent years [1–4]. Soft computing and artificial intelligence are successfully used for machinery control and robotic and engineering applications [5–7]. The integration of nonlinear system with communication technology has led smart and secure industry and home become reality. Therefore, robots may play an important role in such environments.

In addition, indoor inspection, surveillance, and home security are becoming critical issues at this time to organizing a smart and secure place to live and/or work [8, 9]. Advances in closed-circuit TV (CCTV) technology are turning video surveillance equipment into the most valuable loss prevention tool. Such technology can be considered as a safe and secure tool, which is nowadays available for either industrial, commercial or residential applications. The use of inspection and surveillance systems can alert users for threatening situations worsen, as well as for providing an important record of events, such as inspection of a production plants or buildings, for verification of structural and/or electrical components. However, in most of the conventional systems for surveillance, only fixed cameras and fixed infrared cameras are used. Therefore, drawbacks are related to problems in viewing all video streams at the same time or tracking a moving object through a mobile device due to the object dynamics and limited network bandwidth. In order to overcome these limitations, an increasing number of systems have been developed for automatic inspection; they are equipped with sensors allowing the exploration of a building, as reported in [10].

Inspection and monitoring systems are apparently the only tools that are easy to use and manage; in fact, they hide drawbacks such as high purchase and maintenance costs as well as significant financial commitment related to data management and processing. In addition, interoperability and integration with other devices could be a problem; those factors may greatly influence the wide spreading of those systems. In order to enhance the use of these technologies, new solutions have been explored dealing with the concept of robotic and automatic survey using low-cost technology [11]. More specifically, the use of a robotic platform may drastically reduce the time and cost needed for a relief, if compared to a classical approach. Moreover, the use of a low-cost technology, both for the mechanical design of the mobile robot and the onboard sensors, allows the wide spreading of the robotic system and substitution in the case of damages or if the robot is lost. In addition, the developed technology may allow the interoperability of the system and integration in the industrial environment by taking advantage of Industry 4.0.

2. Requirements and solutions for surveillance and inspection robotic systems within Industry 4.0

Surveillance and automatic inspection tasks require a careful analysis to identify basic requirements and appropriate solutions, as it is represented schematically in **Figure 1**. A first basic requirement is related to the mobility issue, which deals with the site to inspect, that is, buildings, industrial plants, or other indoor environment. Indoor and outdoor inspections often require different solutions, either ground or aerial ones. They have in common the need to have a relatively small sized mobile robot that can travel across a large variety of scenarios.

Mobile robots can be classified according to their type of locomotion. Walking systems are well suited for unstructured environment because they can ensure stability and adaptability to a wide range of situations, but they are mechanically

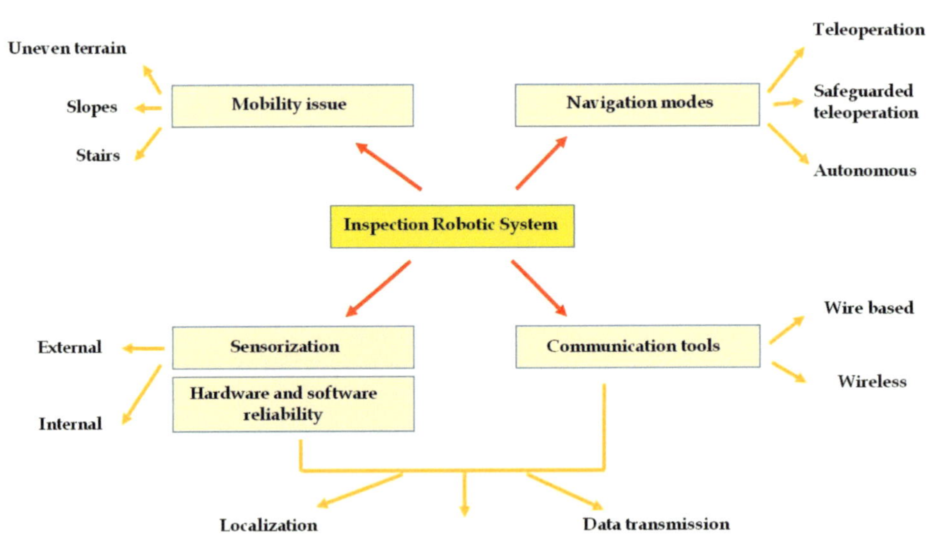

Figure 1.
Flowchart of the main requirements and solutions for a mechatronic design of inspection robotic systems.

complex requiring high power and control efforts [12]. Wheeled locomotion instead is the most efficient solution on flat surfaces [13]. In fact, wheeled or tracked robots are the optimal solution for well-structured environment and quite regular terrain. In off-road, their mobility is often very limited and highly depends on the type of environment and the dimension of the obstacles. Hybrid mobile robots have been developed to combine advantages of legged- and wheeled-/tracked-locomotion types; therefore, lately they are preferred for a large variety of scenarios and applications. Very often the mechanisms used for the locomotion (legs, articulated wheels) should be synthetized ad hoc [14, 15]. A proper model of the mechanical system has to be developed for programming and communication issues [16, 17].

A classification of three types can be made referring to navigation modes, namely, pure teleoperation, safeguarded teleoperation, and autonomous navigation, according to the task and overall budget. The choice depends on the application and the environment. The sensorization is strictly related to the navigation type and the level of sophistication of the inspection. It is possible to classify sensors as internal and external ones. The internal ones give the robot mobility control and navigation capabilities. They can be proximity sensors, encoders, GPS, accelerometers, gyroscopes, magnetic compasses, and tilt and shock sensors. External sensors are related to a specific task; for the inspection and surveillance application, the sensors considered can be cameras, thermal cameras, laser, light, temperature, gas, smoke, oxygen, humidity, sound, and ultrasound. Hardware and software reliability deals with the end user/application of the robot. In fact, this issue has to be set at very early stage of the design process since it is related to operation, maintenance, failure prevention, and intervention.

Communication tools are essential for localization, navigation, or data transmission [16, 17]. In addition to wired and wireless communication, a great challenge is the interoperability with any automatic and robotic system with other devices, such as home automation and security system in industrial plants according to Industry 4.0. This issue is specifically related to the application for surveillance, inspection, and maintenance for structures and infrastructures and in industrial environment, also related to sustainability [18, 19]. Recently, the term Industry 5.0 has been introduced in 2015; comparing with Industry 4.0, which is being considered as the latest

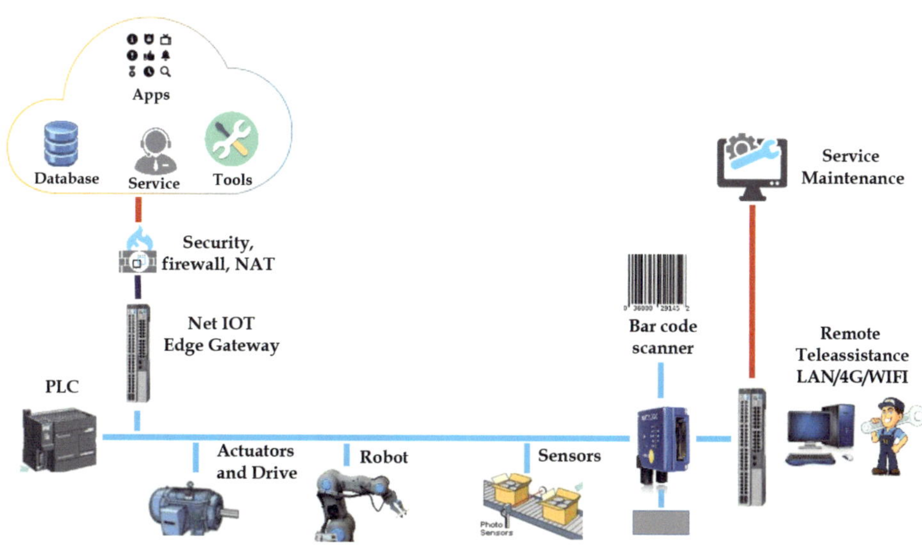

Figure 2.
A scheme for the levels in Industry 4.0 organization and interoperability of robotic systems in industrial environment.

industrial revolution, it can be considered a systemic transformation that includes impact on civil society, governance structures, and human identity in addition to solely economic/manufacturing ramifications, the Industry 5.0 founders prefer to speak about next step in evolution.

The fourth industrial revolution has been applied to significant technological developments several times over the last 75 years and is up for an academic debate. Industry 4.0, on the other hand, focuses on manufacturing specifically in the current context and thus is separate from the fourth industrial revolution in terms of scope. It is very interesting to analyze the significant trimming of the time needed to go from one revolution to the next one. From this regard, the introduction of Industry 5.0 just 4 years after the start of Industry 4.0 is not an exception, but a winner (**Figure 2**).

3. Mechatronic design of inspection robotic system

The mechatronic design of an inspection robotic system can be schematically composed by two main parts: one is related to the robot mobility and operation modes and the other one is responsible to manage the external sensor kit, specifically made according to the application. **Figure 3** shows the mechatronic architecture of the THROO system, proposed by Rea and Ottaviano [10] and used here as paradigmatic example.

According to the scheme reported in **Figure 3**, a tablet in (1) is used for the robot motion control and navigation, taking advantage of the two sliders represented in the zoomed view in (7). The Dension WiRC software is used for programming. The USB WiFi router type is TP-Link Model TL-WN821N, and it is represented in (2) and allows the tablet to access and connect to the WIRC hardware. The webcam (3) is the Logitech U0024 type. The Dension WiRC hardware (4), with four digital inputs, four digital outputs, and eight channels, is used to control the servomotors of the robot. The hardware in (4) is connected to Arduino board (5) and gives the command to the robot's actuators via relay (6). The target (8) is displayed on the tablet.

The overall mechatronic system architecture for the robot navigation is given in (9). For the proposed application (indoor survey) teleoperation mode is used;

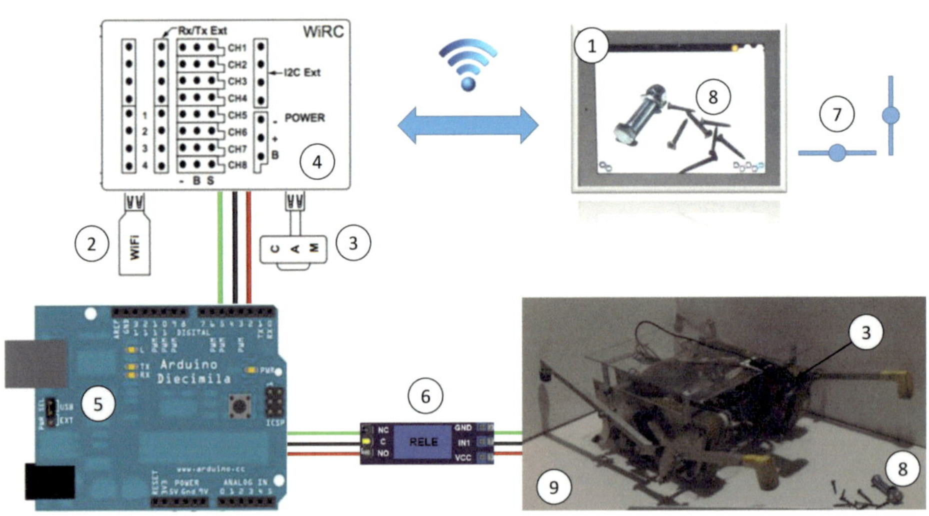

Figure 3.
Mechatronic architecture of the control for the THROO robot.

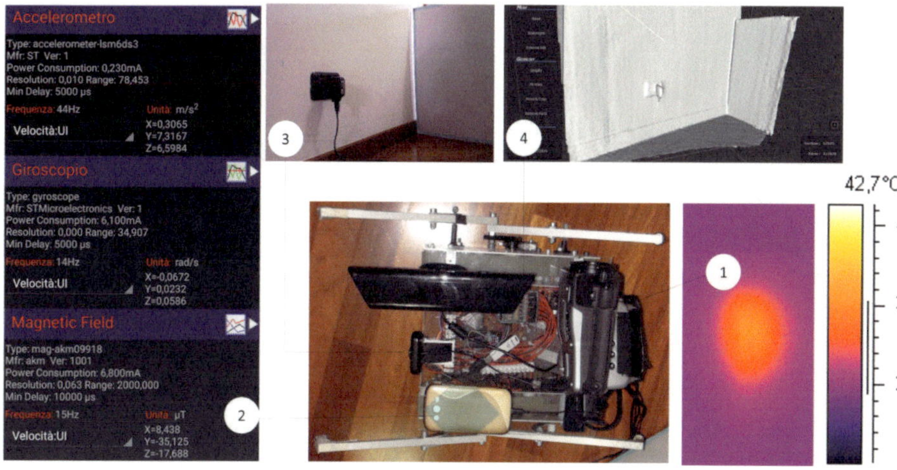

Figure 4.
Sensor installation on the inspection robot.

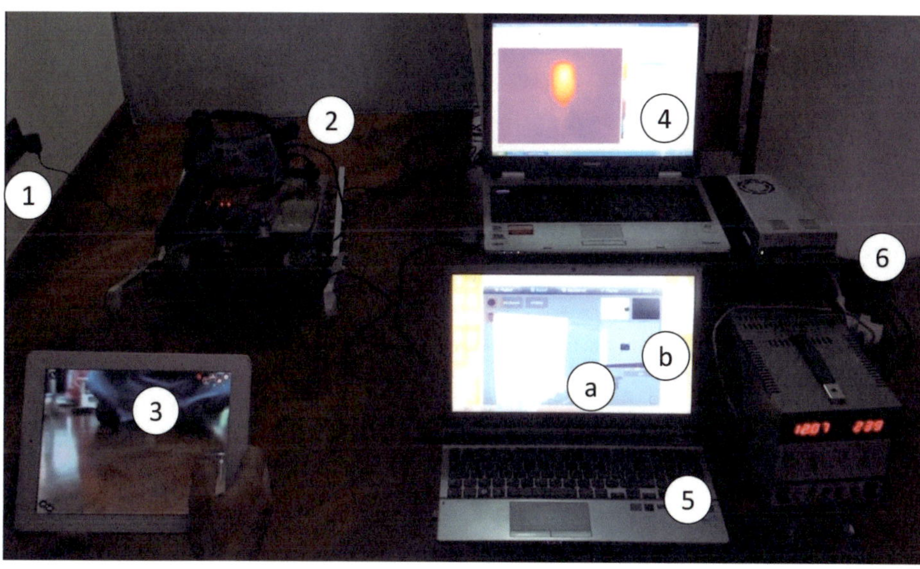

Figure 5.
Overview of the proposed mechatronic/robotic system.

therefore, when the robot in **Figure 4** moves, the internal sensor suite is used to help the user to understand the environment and guide the robot to the path or a close target. In particular, in **Figure 4**, label (2) is for the electronic board equipped with accelerometer, gravity and gyroscope sensors, GPS sensor, magnetic field, and acceleration sensors. The front camera view for navigation is displayed as (3). External sensors are used for inspection and monitoring tasks. More specifically, in **Figure 4**, label (1) is used for the thermal infrared camera FLIR ThermaCAM S40, and the 3D scan is the Xbox Kinect shown in (4), provided with an infrared sensor and two additional micro cameras.

Examples and applications of industrial and nonindustrial applications are reported in [20, 21]. An overall layout for the system is given in **Figure 5**, in which the main components may be recognized, mainly the robot in (2) with external and internal sensors, operating (3) and monitoring (5) (6) systems, and the power

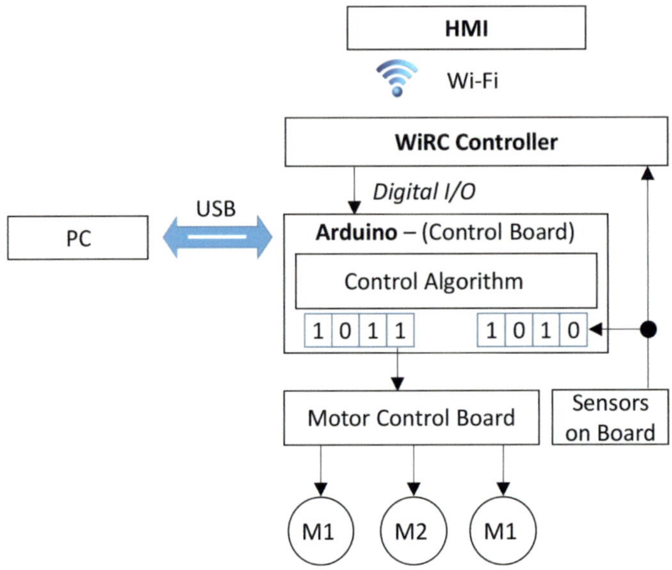

Figure 6.
A scheme for the control architecture.

supply (6); label (1) represents the target. **Figure 6** shows the control architecture to operate the robot and a representation of the interoperation with the equipment onboard.

The use of this kind of control system solution offers several advantages, which can be summarized as follows:

1. Low cost—Arduino boards are relatively low cost, if compared to other microcontroller platforms. They may enable diffusion and affordable cost of the overall system.

2. The Python software runs on Windows, Macintosh OS X, and Linux operating systems. Most microcontroller systems are limited to Windows.

3. Simple, clear programming environment—the programming environment is easy to use for beginners, yet flexible enough for advanced users to take advantage of as well.

4. Open-source and extensible software—the software is made available as open source tools, suitable for extension by experienced programmers. The language can be expanded through C++ libraries.

5. Open-source and extensible hardware—the plans for the modules are published under a Creative Commons license, so experienced circuit designers can make their own version of the module, extending it and improving it. Even relatively inexperienced users can build the breadboard version of the module in order to understand how it works and save money.

CANopen is a CAN-based communication system. It comprises higher-layer protocols and profile specifications. CANopen has been developed as a standardized embedded network with highly flexible configuration capabilities. It was designed originally for motion-oriented machine control systems, such as handling systems.

Parameter description		Specification	
Robot system			
Hybrid mobile robot THROO	Size (L × H × W)	300 (550) × 140 × 400 mm	
	Mass	4.5 kg no batteries	
	Max speed	Up to 0.5 m/s	
	Actuation	DC 24 V 12 Nm 24 W	
	DOFs	2 (track), 1 (legs)	
	Max step size	100 mm	
Internal sensors	*Range*	*Resolution*	*Power*
PSH accelerometer (Intel Inc.)	0–39.227	0.01 (0.024%)	0.006 mA
PSH gyroscope sensor (Intel Inc.)	0–34.907	0.002 (0.005%)	6.1 mA
PSH gravity sensor (Intel Inc.)	0–19.613	0.005 (0.024%)	0.006 mA
PSH magn. field sensor (Intel Inc.)	0–800	0.5 (0.062%)	0.1 mA
PSH lin. accel. sensor (Intel Inc.)	0–19.613	0.005 (0.024%)	0.006 mA
External sensors		*Model*	
Thermal camera (FLIR)		ThermaCAM S40	
Front camera (Logitech)		U0024 type	
3D scan (Xbox)		Kinect	
Communication			
USB WiFi router		TP-Link Model TL-WN821N	
Control station			
No. monitors		2	
No. computers/CPU		2	

Table 1.
System specifications.

Today, it is used in various application fields, such as medical equipment, off-road vehicles, maritime electronics, railway applications, or building automation. In order to obtain this low-cost control system, the advantages of using Arduino with CANopen for data transmission have been combined. This was possible using a PiCAN 2 breadboard interface card that allows Raspberry to send commands via CANopen.

Table 1 summarizes the main features of the proposed system. The integration of sensors and its management has been a subject of research activity in different domains [3, 4, 22–24]. An industrial network laboratory prototype has been proposed by Leão et al. [25] in which several kits have been implemented.

4. Cases of study

The THROO system has been proposed here as a paradigmatic example of a mechatronic solution for the design. Applications reported here refer to indoor surveys. **Figure 7** shows the robot operation during an indoor inspection. In particular, thermal detection of an electrical component is carried out. The interoperability of the sensors onboard with navigation sensorization is managed by the control board and the WiRC controller.

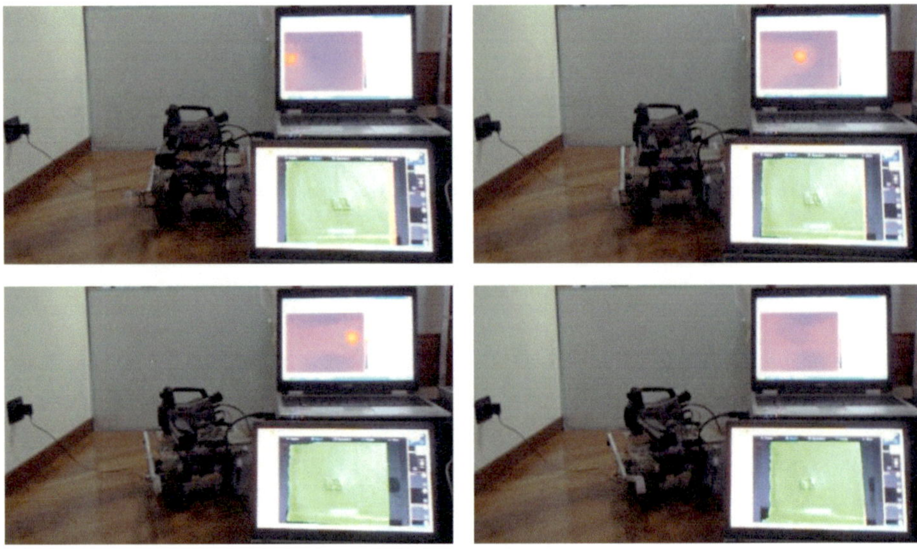

Figure 7.
Photo sequence of an experimental test of data acquisition.

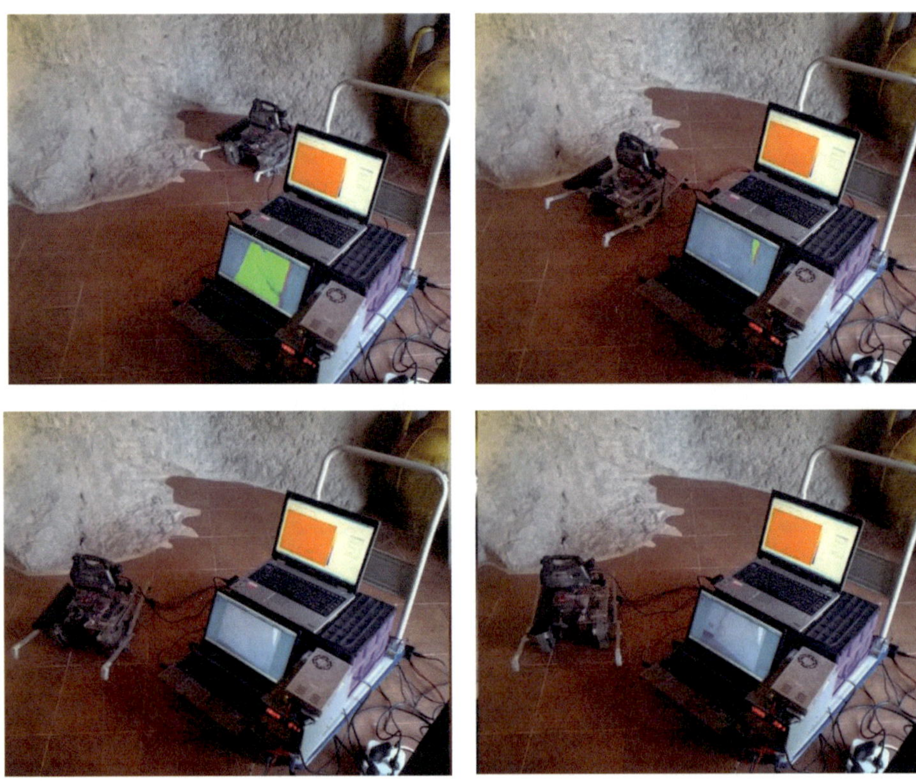

Figure 8.
Photo sequence of data acquisition during a survey.

Figure 8 shows an indoor survey in which an indoor wall element made of limestone rock is analyzed, as reported in [11]. All the acquired data is monitored on two screens as shown in **Figure 8** and stored in a PC for further analysis and reconstruction.

DOI: http://dx.doi.org/10.5772/intechopen.88489

5. Conclusion

In this chapter, we have proposed the main requirements and related items for a mechatronic design of an inspection and surveillance robotic system. The mechatronics and control scheme proposed here might constitute a solution for a broad range of scenarios spacing from home security and inspection of industrial sites, brownfields, historical sites, or sites dangerous or difficult to access by operators. As a paradigmatic example, a hybrid robot is presented here, and experimental tests are reported to show the engineering feasibility of the system and interoperability of the mobile hybrid robot equipped with sensors that allow real-time multiple acquisition and storage. The robot equipment is composed by external and internal sensors, for example, gyroscope, accelerometer, inclinometer, thermal camera, and 3D motion capture system.

Conflict of interest

The authors declare no conflict of interest.

Author details

Pierluigi Rea and Erika Ottaviano*
Department of Civil and Mechanical Engineering, University of Cassino and Southern Lazio, Cassino, FR, Italy

*Address all correspondence to: ottaviano@unicas.it

References

[1] Balaguer C, Montero R, Victores JG, Martínez S, Jardón A. Towards fully automated tunnel inspection: A survey and future trends. In: The 31st International Symposium on Automation and Robotics in Construction and Mining ISARC. Sydney, keynote lecture; 2014. pp. 19-33

[2] Castelli G, Ottaviano E, Rea P. A Cartesian cable-suspended robot for improving end-users' mobility in an urban environment. Robotics and Computer-Integrated Manufacturing. 2014;**30**(3):335-343

[3] Rea P, Ottaviano E, Conte M, D'Aguanno A, De Carolis D. The design of a novel tilt seat for inversion therapy. International Journal of Imaging and Robotics. 2013;**11**(3):1-10

[4] Rea P, Ottaviano E, Castelli G. A procedure for the design of novel assisting devices for the sit-to-stand. Journal of Bionic Engineering. 2013;**10**(4):488-496. ISSN: 1672-6529

[5] Azizi A. Modern manufacturing. In: Applications of Artificial Intelligence Techniques in Industry. Springer Briefs in Applied Sciences and Technology. Singapore: Springer; 2019. pp. 7-17. DOI: 10.1007/978-981-13-2640-0_2

[6] Azizi A, Ghafoorpoor YP, Hashemipour M. Interactive design of storage unit utilizing virtual reality and ergonomic framework for production optimization in manufacturing industry. International Journal on Interactive Design and Manufacturing. 2019;**13**(1):373-381. Article in press. ISSN: 1955-2513 (Print) 1955-2505 (Online)

[7] Azizi A. Introducing a novel hybrid artificial intelligence algorithm to optimize network of industrial applications in modern manufacturing. Complexity. 2017;**2017**:8728209

[8] Borja R, de la Pinta JR, Álvarez A, Maestre JM. Integration of service robots in the smart home by means of UPnP: A surveillance robot case study. Robotics and Autonomous Systems. 2013;**61**:153-160

[9] Tseng C-C, Lin C-L, Shih B-Y, Chen C-Y. SIP-enabled surveillance patrol robot. Robotics and Computer-Integrated Manufacturing. 2013;**29**:394-399

[10] Rea P, Ottaviano E. Design and development of an inspection robotic system for indoor applications. Robotics and Computer-Integrated Manufacturing. 2018;**49**:143-151

[11] Rea P, Pelliccio A, Ottaviano E, Saccucci M. The heritage management and preservation using the mechatronic survey. International Journal of Architectural Heritage. 2017;**11**(8):1121-1132

[12] Figliolini G, Rea P, Conte M. Mechanical design of a novel biped climbing and walking robot. In: ROMANSY 2010, 18th CISM-IFToMM Symposium on Robot Design, Dynamics, and Control. Udine, Italy; 2010. pp. 199-206

[13] Siegwart R, Nourbakhsh IR. Introduction to Autonomous Mobile Robots. Cambridge, Massachusetts: A Bradford Book, The MIT Press; 2004

[14] Borràs J, Thomas F, Ottaviano E, Ceccarelli M. A reconfigurable 5-DoF 5-SPU parallel platform. In: Jian S Dai, Matteo Zoppi, Xianwen Kong editors. Proc. of the ASME/IFToMM Int. Conf. on Reconfigurable Mechanisms and Robots. London, UK: Springer; 2009. Article no. 5173892. pp. 617-623

[15] Figliolini G, Rea P, Angeles J. The synthesis of the axodes of RCCC linkages. Journal of Mechanisms and Robotics. 2016;**8**(2):021011. DOI: 10.1115/1.4031950. ISSN: 19424302

[16] Azizi A, Yazdi PG. Mechanical structures: Mathematical modeling. In: Computer-Based Analysis of the Stochastic Stability of Mechanical Structures Driven by White and Colored Noise. Singapore: Springer; 2019. pp. 37-59

[17] Azizi A, Yazdi PG. Modeling and control of the effect of the noise on the mechanical structures. In: Computer-Based Analysis of the Stochastic Stability of Mechanical Structures Driven by White and Colored Noise. Singapore: Springer; 2019. pp. 75-93

[18] Ghafoorpoor PG, Yazdi P, Azizi A, Hashemipour M. A hybrid methodology for validation of optimization solutions effects on manufacturing sustainability with time study and simulation approach for SMEs. Sustainability. 2019;**11**(5):1454

[19] Yazdi PG, Azizi A, Hashemipour M. An empirical investigation of the relationship between overall equipment efficiency (OEE) and manufacturing sustainability in industry 4.0 with time study approach. Sustainability (Switzerland). 2018;**10**(9):3031

[20] Figliolini G, Rea P. Overall design of Ca.U.M.Ha. Robotic hand for harvesting horticulture products. Robotica. 2006;**24**(3):329-331

[21] Ottaviano E, Ceccarelli M, Castelli G. Experimental results of a 3-DOF parallel manipulator as an earthquake motion simulator. In: Proc. of the ASME Design Engineering Technical Conference; Salt Lake City: Cambridge University Press; Code 64323 2004. Vol. 2A, pp. 215-222. ISSN: 0263-5747. (in print)

[22] Figliolini G, Rea P. Ca.U.M.Ha. robotic hand (cassino-underactuated-multifinger-hand). In: IEEE/ASME International Conference on Advanced Intelligent Mechatronics; Zurich: AIM; 2007. Article number 4412562

[23] Thomas F, Ottaviano E, Ros L, Ceccarelli M. Performance analysis of a 3-2-1 pose estimation device. IEEE Transactions on Robotics. 2005;**21**(3):288-297

[24] Sorli M, Figliolini G, Pastorelli S, Rea P. Experimental identification and validation of a pneumatic positioning servo-system. Power Transmission and Motion Control, Bath, PTCM. 2005;**2005**:365-378

[25] Leão CP, Soares FO, Machado JM, Seabra E, Rodrigues H. Design and development of an industrial network laboratory. International Journal of Emerging Technologies in Learning (iJET). 2011;**6**(2):21-26

Chapter 4

Interaction of Mechatronic Modules in Distributed Technological Installations

Valery A. Kokovin

Abstract

The article deals with the interaction of mechatronic devices in real time through events and messages. The interaction of distributed network devices is necessary to coordinate their work, including synchronization when implementing a distributed algorithm. The approach in the development of a distributed control system (DCS) for mechatronic devices based on the IEC 61499 standard has been analyzed. Using only a LAN for interaction purposes is not always justified, since messages transmitted over a LAN do not provide transmission determinism. To eliminate this problem, a fast local network is needed, which would not utilize resources of the main computer and hardware (e.g., based on the model of terminal machines) to carry out a network communication. It is proposed to implement LAN controllers on the field-programmable gate array (FPGA) platform. Data-strobe coding (DS coding) with a signal level of LVDS was used for keeping the transmitted data intact and improving the overall reliability of the systems.

Keywords: real-time interaction, mechatronic devices, IEC 61499 standard, distributed control systems, field-programmable gate array, a distributed algorithm

1. Introduction

Technological process automation of large industrial or scientific complexes, where there is a large territorial and algorithmic distribution, is related to the development of network systems that provide interaction of individual technological installations. Usually, a telecommunications system acts as the abovementioned network system. In the general case, each installation executes its own part of a given algorithm, operating in an autonomous mode, after which it sends results to other installations. Algorithm execution can result in some informational data, materials, products, etc. A majority of industrial complexes work according to these technologies. Another case is when the result of a distributed technological system work is created by majority of technological subsystems together in real time and the technological processes are related.

The development of technology process management systems is related to building a model through input parameters and current process state parameter formalization. Real-time management of linked, territorial, and algorithmically distributed technology systems with parallel processes is pretty challenging. Additional difficulties appear when linked mechatronic devices that form mechatronic systems are used as an executive object.

Mechatronic components (MCs) can be defined as devices that combine precise mechanical units with electronic computing and management components, interface, and power modules. Such combination allows implementing new features that extend existing functions of the device. At the same time, all mechatronic units work on general task defined for all units. Over the past decades, the MC definition has significantly expanded, and mechatronics became an interdisciplinary industry that can include such disciplines as telecom, robotics, power electronics, etc.

MCs are most commonly used in industries where there are requirements like executing mechanisms' precise positioning, computers' fast response on internal and external events, increased reliability, and limited dimensions.

Very often electronic control components of MC are called embedded control system, which assumes completeness and self-sufficiency of these systems and uses software for running it. Modern MUs that are used in distributed systems have high-performance computers that work within OS environment with support of all common network protocols.

At this moment, MUs and MCs are commonly used as distributed process system (DPS) automation object. All major requirements to MCs as DPS objects are described in article [1]. MCs must have:

- Signal (event) and data interface for interacting with other MDs
- Managed process parameters I/O interface
- Dataset for keeping managed object state
- Manage algorithm for this object.

The chapter is organized as follows. Section 2 provides an analysis of work related to the design of distributed control systems (DCSs) based on the standards IEC 61131-3 [2] and IEC 61499 [3]. Section 3 discusses hardware and software solutions for the implementation of the interaction of mechatronic components. Finally, the conclusion is given in Section 4.

2. Interaction of mechatronic modules in distributed technological systems based on IEC 61131-3 and IEC 61499 standards

2.1 Centralized mechatronic objects control based on standard IEC 61131-3

The IEC 61131-3 [2] standard plays a big role in the automation of technological systems that contain mechatronic blocks. The establishment of this standard allowed unifying development languages of managing applications for programmable logic controller (PLC), which allowed to port developed projects to PLCs manufactured by various vendors. One of graphical languages described by the standard is function block diagram (FBD) language. This language uses the FB concept, which represents part of the program managing code and has an input and output interfaces. The FB interface has special entry calls—event inputs. FBs are combined into chains using interfaces. One of the disadvantages of the FBD language is that a random FB cannot be invoked from a chain.

The PLC software development technology based on the IEC 61131-3 standard is designed for centralized management, which means that direct interaction of separate PLCs is carried through the central computer. This disadvantage makes it significantly more difficult to develop applications for DCSs. When a new PLC is added to the management system or there is a change of PLC interaction algorithm, the central computer program must be changed. Therefore, configuration and scaling of such systems requires sophisticated procedures and takes significant amount of time. Besides that, there is no way to interact directly with distributed technological systems, MCs.

New developments in microelectronics and circuitry allow eliminating some of disadvantages. Thus using a system on a chip (SoC) technology, Altera (Intel) company offered to realize PLC on a crystal, by connecting high-performance dual-core ARM processor and field-programmable gate array (FPGA). This allowed processing of input events in parallel and configuring managing application algorithm remotely [4].

2.2 Distributed control systems on standard IEC 61499

In 2005, a new standard IEC 61499 [4] has been established, which defines a new way of building distributed technological process management systems. The IEC 61499 architecture is based on IEC 61131-3 definitions and uses FBs with extended interface abilities. One of the main FB extensions is an event interface, which allows defining FB execution order explicitly. Each FB can contain several encapsulated algorithms that cannot be accessed from other FBs.

A management system, based on IEC 61499, represents a set of devices that interact with each other using communication network. The management system implements functions described with applications. These applications can be distributed between several devices that can be represented by PLCs, programmable automation controllers (PACs) [5], and digital computers based on FPGA [6] platform. Each device consists of one or several resources. Resource is a functional unit that can manage its operations independently, including algorithm execution. An application is represented by several linked FBs that can be executed on different resources and management system devices.

IEC 614999 FB is an independent program unit that can be created, tested, and used separately from other FBs [3, 7]. The IEC 61499 standard defines three types of blocks: basic function block, service interface function block (mechatronic devices interact using these blocks), and composite function block (contains chain of FBs). **Figure 1** shows the IEC 61499 application model that consists of three FBs combined into one network that is used for events and data flows. FBs can be in certain states, set by function algorithm of the FB. Transfer from one state to another can be triggered by an event, received from a neighbor FB.

The main advantage of designing management systems based on the IEC 61499 standard is simplicity of reconfiguration, flexibility, and reduced development time, reusing developed components and scalability. Article [8] analyzes features of this method of developing distributed technological systems, its advantages and

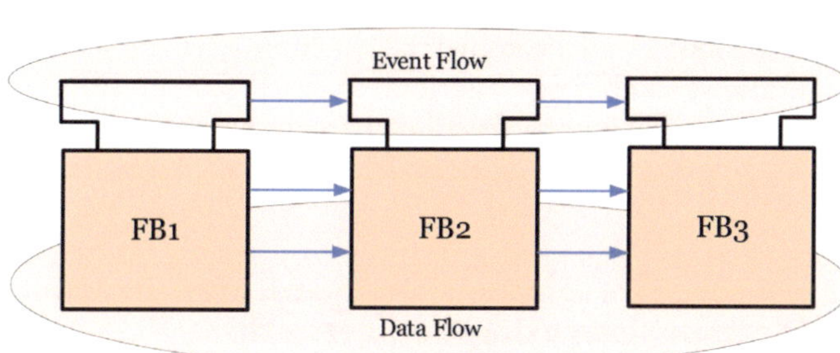

Figure 1.
IEC 61499 application model.

disadvantages. For keeping backward compatibility while moving from developing managing applications based on PLC IEC 61131-3 to applications based on IEC 61499, there is a methodology based on web technologies [9]. The main goal set in that article is to keep the identity of application behavior when launched on the IEC 61499 platform.

The DPS stability is defined by distinctness of executing applications. The IEC 61499 standard defines FB as an abstract model, which allows various FB behavior interpretations. Article [10] analyzes FB execution models. The standards sections related to the base FB semantics describe a situation, when only one FB can be active at any period of time within one network. This limitation allows to develop execution models in two directions: serial and cyclic (by analogy with PLC cyclic processing) models.

However, the evolution of multiprocessor and multicore computers allows running a parallel model. Article [11] contains suggestions about running such models. Besides, a parallel-type model can be implemented on FPGAs with the ability to use real parallelism, which is very important for mechatronic systems that are critical to the response time to incoming events.

Article [12] defines the Intelligent Mechatronic Component (IMC) and describes the conditions of using such components in DPS based on IEC 61499. Any IMC can contain the following elements:

- Be a mechatronic device, thus to represent a physical functional device with sensors, actuators, and electronic circuits

- An integrated control device, which is a computing device that includes interfaces to sensors, actuators, and communication networks, for interaction with other IMCs

- Software with data support and control logic for implementing automation functions of IEC 61499 standard

Most of the works related to the development of automation projects based on IEC 61499 are of research in nature. Despite the obvious advantages of the standard (described earlier) in the design of distributed control systems, widespread implementation in the industry has not yet happened. Article [13] describes problems of the standard that prevent it from being fully used: semantic problems, the lack of well-developed design methodologies, restrictions on the use of various execution models, and others. There is also a lack of integrated design methodologies that facilitate component-based design throughout the design cycle of automation systems [12].

The standard is being developed, and its individual provisions are being clarified. For practical use, the IEC 61499 compliance profile [14] is issued and constantly updated. There are commercial development tools for creating projects that meet the IEC 61499 standard. The most well-known project is ISaGRAF [15], in the form of a design support tool (workbench), which works with both the IEC 61131-3 and IEC 61499 standards. The nxtSTUDIO commercial system project [16] focuses exclusively on IEC 61499 and allows combining a distributed control system with HMI/SCADA. In addition, nxtSTUDIO makes it possible to automate the process of building communication channels between controllers of mechatronic devices of a distributed system.

3. Hardware and software solutions for organizing interaction of mechatronic components using local networks

The development of microelectronics and information technology allows creating miniature computers with low power consumption and high performance, which can be embedded not only in technological systems but also in individual elements of these systems, including mechatronic devices. This feature, as mentioned above, makes it possible to create intelligent subsystems combined by a global or local communication network.

The concept of the Internet of Things (IoT) [17], successfully introduced into production, gave an impetus to the development of a new direction that unites robots or robotic devices using network technologies. This direction, called the Internet of Robotic Things (IoRT), is aimed at implementing robotic technologies, by extending the functionality of IoT devices. In [18], the IoRT concept is presented, which emphasizes the tremendous flexibility in developing and implementing new applications for networked robotics while achieving the goal of providing distributed computing resources as the main utility. In these network associations (IoT and IoRT), the term "Internet" can be interpreted as a global association of computer networks that use TCP/IP protocols when interacting with each other. In addition, the participants of network associations use a huge variety of interfaces and protocols when receiving information from sensors and transmitting control signals to actuators.

By analogy with the network associations of functional devices given above, we can speak about the Internet of Mechatronic Components, a subset of which is IoRT. On the other hand, in distributed technological systems among the participants, there can be not only MCs but also, for example, self-sufficient electrical devices with integrated intelligence, that is, devices that do not include mechanics. Self-sufficiency means the ability of devices to solve independently a part of a distributed technological problem delegated to them, but which need to exchange information with other participants through the formation of events or messages. Conditionally, such devices can be called functional network connectivity (FNC).

3.1 FNC-embedded computer

Article [19] presents a methodology for creating a prototype intelligent power electronic converter (iPEC) with the ability to remotely control and interact with other DPS devices. In the developed iPEC prototype, the power module implements the function of a powerful generator of infra-low-frequency voltage. This generator can be used to power equipment (piezoelectric elements) for ultrasonic cleaning of parts or for solving geophysical tasks (cleaning of wells). An example of iPEC-type device use can be found in article [20]. This article presents solution of the technological problem of the formation of a uniform effect of ultrasonic vibrations in a liquid medium on the processed. To solve the problem, it is proposed to use a whole matrix of distributed ultrasound transducers placed in a certain way. Control signals arriving at the transducers are formed in such a way that areas with low and high pressure of the ultrasonic field arise, which causes the occurrence of directional flows and contributes to the cleaning of products. Drivers of control signals are subject to stringent requirements for response time.

The given example shows the need to create a universal built-in control module, both as part of MCs and as part of other DPS participants, that could provide execution of the following main tasks:

- Global network communication task
- Local network communication task
- Management task for the implementation of the built-in algorithm
- Sensor data processing task

In a distributed technological system, participants called as FNCs can be used as independent technological subsystems, or they can be clustered, by analogy with the composite function block (IEC 61499). On this basis, the interaction of individual FNCs can be carried out over several communication networks—global and local. The global network is available to all exchange participants, and the local network is available only to cluster devices. When interacting over a local network, an additional requirement is added—the determinism of the transmitted events and messages.

The interaction of DPS devices is necessary to coordinate their work, including synchronization when implementing a distributed algorithm. Article [19] justifies the use of the hybrid configuration of the computational control part of FNC-type devices. A device can contain several different computators, based on various calculation models. These can be computators implemented on processors with ARM architecture [21] and on the FPGA platform. Sharing two different computers gives the most flexible functionality.

Important benefits of using ARM controllers in FNC are high performance and low power consumption. This makes it possible to build a universal-embedded computer for both stationary technological systems and mobile ones. For example, in computers based on the Smart Mobility ARChitecture (SMARC) standard [22], ARM controllers are widely used.

ARM controllers used in the FNC solve the following tasks: communication over a global network, downloading computer configurations, implementing an embedded algorithm, implementing cloud technologies, video processing of objects for machine vision, etc. Global network controllers must support TCP/IP protocols, which are typically present in modern embedded computers. The exchange between mobile autonomous objects, for example, automatic guided vehicle (AGV) [23], requires a wireless network controller as part of the network module.

The ideology of random access to the network is not suitable for solving the problem of interaction between FNC-cluster devices, since it is impossible to ensure the determinism of message delivery. Therefore, a fast local network is needed, which would not utilize resources of the microcontroller and hardware (e.g., based on the model of finite automata) to carry out a network exchange. Also, it is necessary to use a special encoding at the signal level when transmitting events and messages, for keeping the transmitted data intact and improving the overall reliability of the systems. Controllers of such a network are usually implemented on the FPGA platform. When creating equipment based on FPGA for technological installations of the accelerator complex [24], the implementation of the exchange based on data-strobe coding (DS coding) network signals with the signal level LVDS [25] showed a good result. LVDS levels have high noise immunity and energy efficiency. DS coding allows you to transfer data at high speeds without first agreeing the speeds between the calculators of the two devices. At a speed of 100 Mb/s, the length of the communication line can be within 30 m. DS coding has been successfully applied in the aerospace industry within the framework of the SpaceWire standard, where increased requirements are applied to the reliability of transmitted data [26]. Another option for organizing a local network can be using the MIL-1553 interface. Article [27] describes a network for transmitting events and messages for technological subsystems of the accelerator complex.

Figure 2 shows an example of the simplified structure of the FNC computer, which includes support for communication over local and global networks. The computer contains:

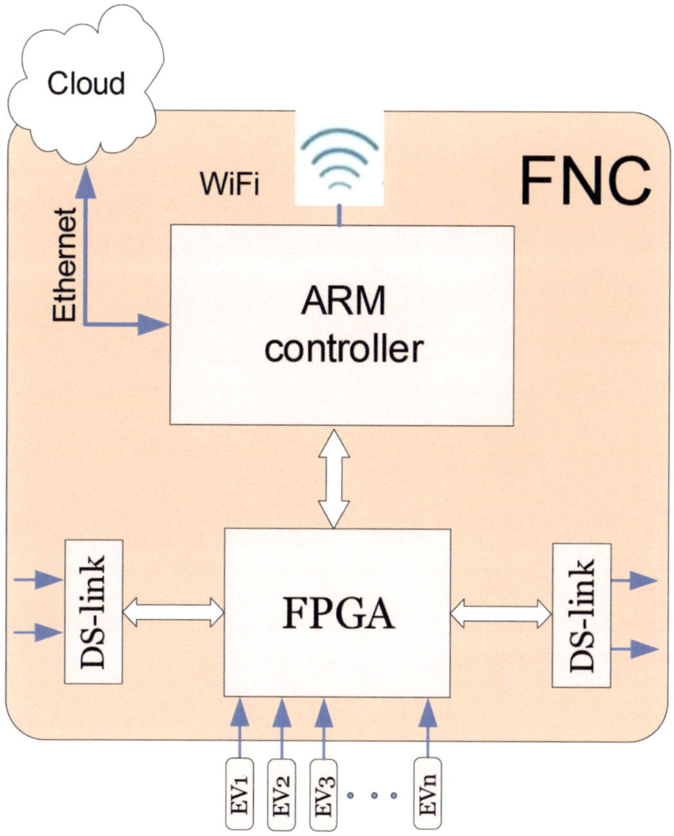

Figure 2.
FNC-embedded computer.

- ARM controller for wired and wireless networking with other members of the global network
- LAN controller on the FPGA platform
- FPGA signal level converters to LVDS levels (DS links).

FPGA FNC-embedded computer receives events (EV1.EVn) that must be transferred to other FNCs of the cluster. **Figure 2** does not show network support for the lower level of the sensors. Thus, we can conclude that, for full-fledged work as part of a distributed technological system, each FNC should have at least three levels of network support.

3.2 Communication network configuration

The choice of the type and number of communication networks in DPS is determined by many factors: the technological problem to be solved, the network configuration chosen, geographical distribution of the interacting subsystems, and so on. Article [28] justifies a circular network configuration consisting of series-connected MCs. Serial network is formed using a single duplex DS link. Messages are transmitted sequentially from one MC to another using a DS interface and can have a broadcast status or contain the address of a particular MC. To control the

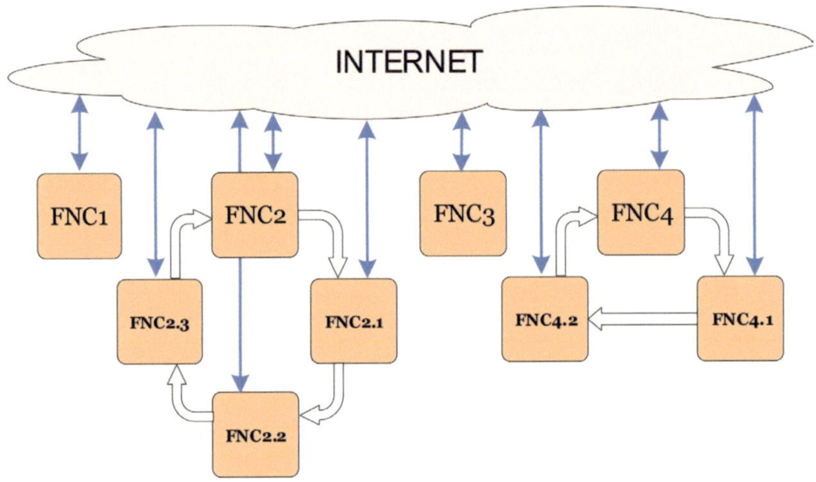

Figure 3.
FNC networks.

passage of messages, the last MC is connected to the first. The advantages of a ring serial network are as follows:

- DS link of each MC is a network repeater and amplifier that allows you to maintain a high speed with a large number of mechatronic devices.
- It is possible to control the transmitted message for accuracy and transmission time upon return after passing through the entire network.
- Message transmission time is strictly determined.

Figure 3 shows the structure of a distributed technological system with two communication networks—global and local. Each FNC device may be a specific functional technological unit. It can be a mechatronic module, individual components of a robotic system, or a power converter. The devices FNC2, FNC2.1, FNC2.2, and FNC2.3 and FNC4, FNC4.1, and FNC4.2 are clustered.

Each FNC device has access to the global network to perform tasks that require access to common databases: reconfiguration, transfer of measured process parameters or video data, and so on. Only FNC1, FNC2, FNC3, and FNC4 exchange event control via the global network. Devices within clusters transmit messages sequentially within the cluster.

4. Conclusion

Intellectualization of mechatronic devices allows expanding the functionality of technological systems. The intelligence of the MCs is provided by the increased computational performance of the controllers, the additional real-time video processing capability, and high sensory sensitivity. Many algorithmic problems that were previously solved with the involvement of a centralized computer can now be solved at the level of the mechatronic component itself. The increased intelligence of the MCs adds increased requirements for organizing the design of distributed control systems, complicating the solution of the problem of interaction of distributed technological subsystems. A rather long implementation of the IEC 61499 standard showed the complexity of the problem to be solved. The development of mechatronic components as network devices goes toward the unification of embedded computers on the proposed computing and communication services. Standards for the design of distributed control systems are being implemented, which determine the ideology of the interaction of a distributed algorithm of a technological problem.

Author details

Valery A. Kokovin
State University "Dubna" Branch "Protvino", Moscow region, Russia

*Address all correspondence to: kokovin@uni-protvino.ru

References

[1] Panjaitan S, Frey G. Functional control objects in distributed automation systems. IFAC Proceedings Volumes. 2007;**40**(3):259-264

[2] International Standard IEC 61131-3 (edition 2.0): Programmable Controllers/ International Electrotechnical Commission; Geneva; 2003. p. 230

[3] International Standard IEC 61499. Function blocks for industrial-process measurement and control systems. Part 1: Architecture/International Electrotechnical Commission; Geneva; 2005. p. 245

[4] Integrating PLC Systems on a Single FPGA or SoC. Available from: https://www.intel.com/content/dam/www/programmable/us/en/pdfs/literature/po/ss-plc-on-a-single-chip.pdf [Accessed: 28 May 2019]

[5] Programmable Automation Controller. Available from: http://www.cannonautomata-products.com/programmable-automation-controller.html [Accessed: 28 May 2019]

[6] O'Sullivan D, Heffernan D. VHDL architecture for IEC 61499 function blocks. IET Computers and Digital Techniques. 2010;**4**(6):515-524

[7] Dubinin V, Vyatkin V. On definition of a formal model for IEC 61499 function blocks. EURASIP Journal Embedded Systems. 2008;**2008**:1-10

[8] Vyatkin V. IEC 61499 as enabler of distributed and intelligent automation: State of the art review. IEEE Transactions on Industrial Informatics. 2011;7(4):768-781

[9] Dai W, Dubinin VN, Vyatkin V. Migration from PLC to IEC 61499 using semantic web technologies. IEEE Transactions on Systems, Man, and Cybernetics. 2013;**44**(3):277-291

[10] Vyatkin V. The IEC 61499 standard and its semantics. IEEE Industrial Electronics Magazine. 2009;**3**:40-48

[11] Vyatkin V, Dubinin V, Ferrarini LM, Veber C. Alternatives for execution semantics of IEC61499. In: 5th IEEE Conference on Industrial Informatics; Vienna; 2007. pp. 1151-1156

[12] Pang C, Vyatkin V. IEC 61499 function block implementation of intelligent mechatronic component. In: 8th IEEE Conference on Industrial Informatics (INDIN 2010); 2010. pp. 1124-1129

[13] Thramboulidis K. IEC 61499 in factory automation. Advances in Computer, Information, and Systems Sciences, and Engineering. 2006: 115-124. Available from: https://doi.org/10.1007/1-4020-5261-8

[14] IEC 61499 Compliance Profile for Feasibility Demonstrations. Available from: https://www.holobloc.com/doc/ita/index.htm [Accessed: 28 May 2019]

[15] ISaGRAF Technology. Available from: https://www.rockwellautomation.com/global/detail.page?pagetitle=Isagraf&content_type=tech_data&docid=209076c017d6dd586c895e9e3a4856e4&redirect_type=tld&redirect_url=www.isagraf.com [Accessed: 28 May 2019]

[16] nxtSTUDIO. Available from: https://www.nxtcontrol.com/en/engineering/ [Accessed: 28 May 2019]

[17] Javed B, Iqbal MW, Abbas H. Internet of things (IoT) design considerations for developers and manufacturers. ICC2017

[18] Ray PP. Internet of robotic things: Concept technologies and challenges. IEEE Access. 2017;**4**:9489-9500

[19] Kokovin V, Diagilev V, Halik J, Uvaysova S. Intelligent power electronic converter for wired and wireless distributed applications. In: Proceedings of the IEEE International Conference SED-2019; 23-24 April 2019; Prague: IEEE; 2019. (in the press)

[20] Dostanko AP, Avakov SM, Ageev OA. Tekhnologicheskie Kompleksy Integrirovannykh Protsessov Proizvodstva Izdeliy Elektroniki [Technological Systems Integrated Production Processes of Electronics Products]. Belorusskaya nauka: Minsk; 2016. pp. 5-25

[21] Available from: https://www.arm.com/ [Accessed: 28 May 2019]

[22] Smart Mobility ARChitecture Hardware Specification. Available from: https://sget.org/standards/smarc/ [Accessed: 28 May 2019]

[23] Available from: http://www.roboticautomation.com.au/agvs [Accessed: 28 May 2019]

[24] Kalinin AY, Kokovin VA, Kryshkin VI, Skvortsov VV. An absolute intensity beam monitor. Instruments and Experimental Techniques. 2016;**59**(4):536-358

[25] ANSI/TIA/EIA-644-1995. Electrical characteristics of low voltage differential signaling (LVDS) interface circuits; 1995

[26] SpaceWire Standard. ECSS—Space Engineering. "SpaceWire—Links, nodes, routers and networks." ECSS-E-ST-50-12C; Rev1 Draft D; November 2014

[27] Komarov V, Antonichev G, Kim L, Kokovin V, et al. Modernization of U-70 general timing system. In: Proceedings of ICALEPCS-2005; Geneva, Switzerland; 10-14 October 2005

[28] Kokovin VA, Evsikov AA. Event-related interaction of mechatronic modules in distributed technological installations. Mechanical Engineering Research and Education. 2018;**5**:12

Chapter 5

Impact Analysis of MR-Laminated Composite Structures

Abolghassem Zabihollah, Jalil Naji and Shahin Zareie

Abstract

Laminated composite structures are being used in many applications, including aerospace, automobiles, and civil engineering applications, due to their high stiffness to weight ratio. However, composite structures suffer from low ductility and sufficient flexibility to resist against dynamic, particularly impact loadings. Recently, a new generation of laminated composite structures has been developed in which some layers have been filled fully or partially with magnetorheological (MR) fluids; hereafter we call them *MR-laminated structures*. The present article investigates the effects of MR fluid layers on vibration characteristics and specifically on impact loadings of the laminated composite beams. Experimental works have been conducted to study the dynamic performance of the MR-laminated beams.

Keywords: laminated composite, MR fluid, vibration, impact

1. Introduction

Laminated composite materials, due to their unique characteristics such as high strength-to-weight ratio, high corrosion and impact resistance, and excellent fatigue strength, are being widely used in aerospace, automobiles, and recently civil engineering applications [1–7]. However, composite structures are highly vulnerable for failure under dynamic loading and, most importantly, impact loads. In the past few years, elements of smart and functional materials have been added to the conventional composite structures to develop a new generation of the laminated composite structures [8]. Researchers have proposed and test many types of smart materials, including piezoelectric, shape memory alloys, fiber optics, and electrorheological (ER) and magnetorheological (MR) fluids, to add the required features, such as controllability, and improve the performance for specific applications [9–14]. These structures have the capability to adapt their response to external stimuli such as load or environmental changes. These new structures have opened new challenges in research communities. The use of MR fluids in composite structures is relatively new as embedding fluids inside a rigid, laminated structure may raise many challenges in fabrication. **Figure 1** shows a typical MR-laminated beam in which some layers are partially replaced by segments of MR fluid. Yalcintas and Dai [15] investigated the dynamic vibration response of three-layered MR and ER adaptive beams both theoretically and experimentally.

Sapiński et al. [16, 17] explored vibration control capabilities of a three-layered cantilever beam with MR fluid and developed FEM model to describe the phenomena in MR fluid layer during transverse vibration of the beam. Sapiński et al. [18] proposed a finite element (FE) model by using ANSYS for sandwich beam

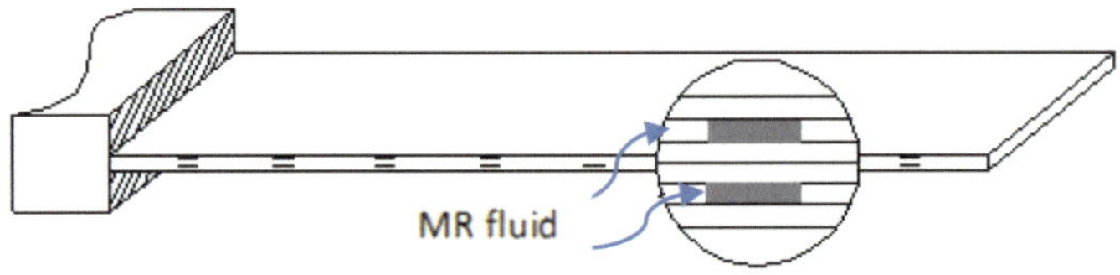

Figure 1.
MR-laminated beam.

incorporating MR fluid. Rajamohan et al. [19] investigated the properties of a three-layer MR beam. The governing equations of MR adaptive beam were formulated in the finite element form and also by Ritz method. Ramamoorthy et al. [20] investigated vibration responses of a partially treated laminated composite plate integrated with MR fluid segment. The governing differential equations of motion for partially treated laminated plate with MR are presented in finite element formulation. Payganeh et al. [21] theoretically investigated free vibrational behavior of a sandwich panel with composite sheets and MR layer. They studied effects of length and width of sheet and also core thickness on frequency. Aguib et al. [22] experimentally and numerically studied the vibrational response of a MR elastomer sandwich beam subjected to harmonic excitation. They studied the effect of the intensity of the current flowing through a magnet coil on several dynamic factors. Naji et al. [23] employed generalized layerwise theory to overcome this challenge and passed behind constant shear deformation assumption in MR layer that is mainly used for vibration analysis of MR beam. Based on layerwise theory, FEM formulation was developed for simulation of MR beam, and results were verified by experimental test. Naji et al. [24, 25] presented a distinctive and innovative formulation for shear modulus of MR fluid. Most recently, Momeni et al. [26, 27] developed a finite element model to investigate the vibration response of MR-laminated beams with multiple MR layers through the thickness of the laminated beam with uniform and tapered cross sections.

The present work intends to study the vibration response of MR-laminated beam with emphasis on impact loadings. The mathematical modeling of MR-laminated beam is similar to the work done by present authors in previous publications [23, 24, 26, 27]. However, here, the modeling has been used mainly to study the effects of impact loading, although for more clarification, basic results for natural vibration have also been provided. Some experimental works have been conducted to illustrate the performance of the MR-laminated beam under practical impact loadings.

2. Review the modeling laminated composite beams with layerwise displacement theory

In brief, laminated composite plates are composed of individual layers, which have been stacked together, usually by hand-layup techniques. Individual layers are composed of fibers, which have been derationed according to property requirement, and matrix, which serves as binder of fibers and transfers the loads to the fibers. Changing the orientation of the fibers optimizes the composite material for strength, stiffness, fatigue, heat, and moisture resistance. Modeling laminated composite structures for conventional applications mainly is conducted by considering the stacked layers as one single layer. The equivalent single-layer (ESL) theories assume continuous displacement through the thickness of the laminate. In general, the stiffness of the adjacent layers in the laminates is not equal; thus, it results in discontinuity in transverse stress through the thickness, which is contrary to the equilibrium of the interlaminar stresses as stated by ESL. In general, ESL theories provide acceptable results for relatively thin laminate. For thick laminate and laminate with material and/or geometric inhomogeneities, such as MR-laminated beams, the ESL theories lead to erroneous results for all stresses.

In smart-laminated structures, due to the material and geometric inhomogeneities through thickness, including MR-laminated beams, it is required to acquire an accurate evaluation of strain–stress at the ply level. Interlaminar stresses can lead to delamination and failure of the laminate at loads that are much lower than the failure strength predicted by the ESL theories.

The accurate modeling of interlaminar stress field in composite laminates requires the displacement field to be piecewise continuous through the thickness direction. Researchers in composite communities have proposed and developed a variety of displacement models to provide sufficient accuracy for interlaminar stresses in composite structures. Layerwise displacement theory developed by Reddy [28] developed a layerwise theory based on the piecewise displacement through the laminate thickness.

The layerwise formulation has the capability to address local through-the-thickness effect, such as the evolution of complicated stress–strain fields in MR-laminated composite structures and interfacial phenomena between the different embedded layers. In this work, the layerwise displacement theory has been used for modeling MR-laminated beam.

The displacement field for a laminated beam based on the layerwise theory is obtained by considering the axial and through-the-thickness displacements as

$$u(x,z,t) = \sum_{I=1}^{N} U_I(x,t)\,\Phi^I(z), w(x,z,t) = \sum_{I=1}^{N} W_I(x,t)\,\Phi^I(z)\;\Phi^I(z) = [1-\zeta\;\;\zeta] \quad (1)$$

where u and w are displacements along x- and z-directions, respectively. N denotes the total number of nodes through the thickness. The ratio ζ is defined as $\zeta = z/h$, in which h represents the thickness of each discrete layer. Interpolation functions $\Phi(z)$ are defined between any two adjacent layers. For thin laminate, displacements in z-direction between layers are negligible, so $w(x,z,t) = W_o(x,t)$.

2.1 Finite element formulation

For many applications, closed-form solution for MR-laminated beams is either not available or very complex. Therefore, most of the computations are based on finite element models. Finite element formulation has been obtained by incorporating the local in-plane approximations for the state variables introduced in Eq. (1) as follows:

$$U_I = \sum_{i=1}^{Nn} U_I^i(x)\,\phi_i(x) \quad (2)$$

where N_n is the number of nodes and $\phi_i(x)$ are the interpolation functions along the length of the beam, respectively. For details on finite element modeling of composite structures based on layerwise theory, one may refer to the work done by the present author in Ref. [8].

3. Fundamentals of magnetorheological (MR) fluids

Magnetorheological (MR) fluid is a class of new intelligent materials, which rheological characteristics such as the viscosity, elasticity and plasticity change rapidly (in order of milliseconds) subject to the applied magnetic field as shown in **Figure 2**. By applying a magnetic field, the particles create columnar structures

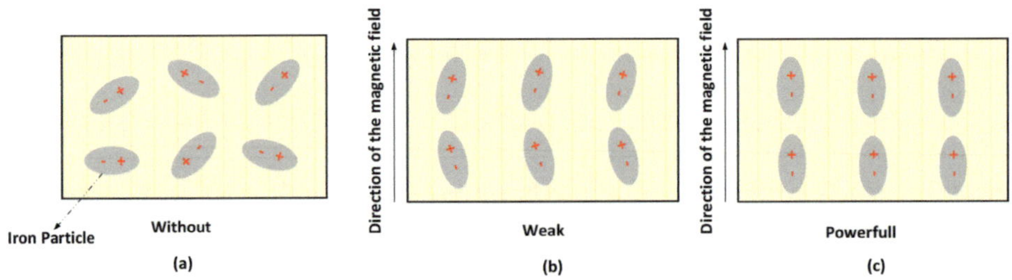

Figure 2.
Structure of MR fluids (a) without magnetic field, (b) with magnetic field, and (c) with magnetic field and shear force.

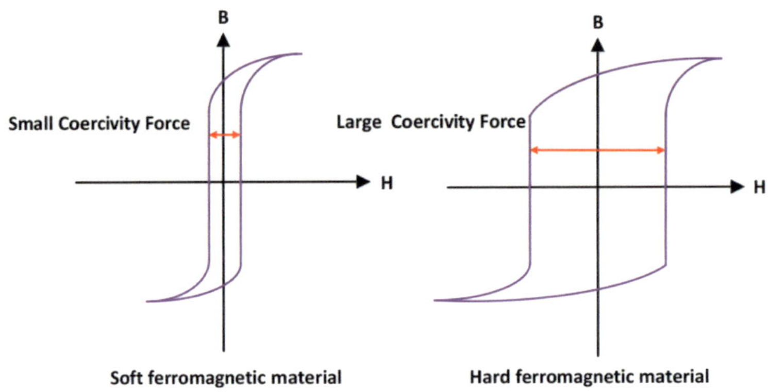

Figure 3.
The comparison between the soft and hard ferromagnetic materials [33].

parallel to the applied field and these chain-like structures restrict the flow of the fluid, requiring minimum shear stress for the flow to be initiated. Upon removing the magnetic field, the fluid returns to its original status, very fast.

Overall, MRF is composed of a carrier fluid, such as silicone oil, and iron particles, which are dispersed in the fluid [29, 30]. Each of the components plays a significant role in the characteristic of MRF.

3.1 Ferromagnetic particles

There are two types of ferromagnetic materials, used in MRFs: the soft material and the hard material [31]. The applied magnetic field intensity (H) and the magnetization of the material (B) are two parameters that show the difference between the two kinds of materials.

Figure 3 shows the H-B hysteresis loops of the soft and hard materials. It is noted that the soft ferromagnetic material has lower remanence and coercivity [32].

The shape of the magneto-soft particles is spherical with a diameter ranging between 1 and 10 μm. One of the most widely used soft materials is carbonyl iron, and that can take up to 50% of the volume of MRF [32]. In order to improve the MRF's dynamic behavior, iron-cobalt and iron-nickel alloys can be used instead of conventional carbonyl iron particles. However, they are more expensive [32].

The hard magnetic material can be made of chromium dioxide (CrO_2), which shows high coercivity and remanence. The size of dispersed CrO_2 is between 0.1 and 10 μm, and it can be easily oriented under the influence of magnetic fields [32].

Commercial MRF	Percent iron by volume	Carrier fluid	Density (g/cm^3)
MRF-122-2ES	22	Hydrocarbon oil	2.38
MRF-132 AD	32	Hydrocarbon oil	3.09
MRF-336AG	36	Silicone oil	3.45
MRF-241ES	41	Water	3.86

Table 1.
Properties of commercial MRFs [34].

It mixes with the soft magnetic material to enhance the rheological behavior and provide more stability to the MRF [32].

3.2 Carrier liquids

The second component of MRF is a carrier liquid providing the continuous medium for the ferromagnetic particles [32]. Although all types of fluids are suitable for this purpose, Ashour et al. [13] recommended a fluid with a viscosity ranging between 0.01 and 1 N/m^2 [32]. Silicon oil and synthetic oil are samples of suitable carrier fluids [32]. **Table 1** provides the samples of carrier fluids with their specifications. As noted in **Table 1**, water is the suitable fluid, which can carry iron particles up to 41% of volume. Hydrocarbon oil and silicone can have the suspended particles between 22 and 36% of the volume.

3.3 Stabilizers

The third element of the MRF is a stabilizer, which are "polymers (surfactants) in nonpolar media." The main roles of the materials as stabilizer are to retain the iron alloy particles suspended in the MRF, extend the service life of the smart fluid, and increase its reliability [32, 35]. There are three types of stabilizers—the agglomerative, the sedimental, and the thermal [32]—as briefly described below:

1. An agglomerative stabilizer is used to prevent the formation of aggregates between the iron particles. Agglomeration usually occurs due to the van der Waals' interactions between the iron particles in the MRF, causing the ferromagnetic particles to stick together. In MRF applications, surfactants should be selected in accordance with the type and concentration of particles. In fine-dispersed concentrations, when the iron particles fill up to 50% of the volume, ionic or nonionic surfactants are recommended. In lower concentrations of iron particles, which only occupy up to 10% of the volume, gel-like stabilizers are also suggested [32].

2. A sedimental stabilizer is employed to prohibit the iron particles from settling down as a result of gravity which decreases the effectiveness of MRFs [32]. The gel-forming and nonionic surfactants, as the sedimental stabilizer, are used to be added to the fluid carrier [35].

3. A thermal stabilizer is utilized to stabilize MRF over a wide temperature range particularly for the long-term applications at high temperature.

Since then, MR fluids have been utilized in various applications, including dampers, brakes and clutches, polishing devices, hydraulic valves, seals, and flexible fixtures. Recently, the application of MR fluids in vibration control has been attracted by many researchers. However, due to the nature of MR fluids, it is very difficult to integrate them with thin-laminated composite structures.

4. Governing equation of MR-laminated beams

Using the displacement field given in Eq. (1) and following a finite element procedure for each element, the governing equation of motion of MR-laminated beam in the matrix form is defined as

$$[m^e]\{\ddot{d}\}_i + \left[k^e(B,f)\right]\{d\}_i = \{f^e\} \tag{3}$$

where $[m^e]$ and $[k^e]$ are the element mass and stiffness matrices, respectively, and $\{f^e\}$ is the element force vector. One may note that when using layerwise displacement theory, depending on the number of layers in the laminate, each node may have many degrees of freedom.

Considering axial displacement given in Eq. (2), the generalized equation of motion of the MR-laminated beam can be obtained by assembling the mass, stiffness matrices, and the force vector as

$$[M]\{\ddot{d}\} + \left[K(B,f)\right]\{d\} = \{F\} \tag{4}$$

where $[M]$ and $[K]$ are the MR-laminated beam mass and stiffness matrices, respectively, $\{F\}$ is the force vector, and $[d]$ is the displacement vector.

It should be noted that the stiffness matrix is a complex value since it is the summation of the stiffness matrix of the laminated layer $[K_c]$ and the stiffness matrix of the MR fluid $[K_{MR}(B, f)]$:

$$\left[K(B,f)\right] = [K_c] + \left[K_{MR}(B,f)\right] \tag{5}$$

One important feature in Eq. (4) is the structural damping which is included in the stiffness matrix. For the sake of simplicity, the complex mathematical modeling is neglected here; however, complete details are available in Ref. [23].

5. Experimental setup for vibration and impact tests

In order to conduct experimental test for MR beams, it is essential to provide a uniform magnetic field all over the beam. For the current work, an electromagnet device shown in **Figure 4**, which was previously fabricated by the current authors at Sharif University of Technology, is used. The length of poles (gap) is 240 mm. The space between the poles (gap) is 40 mm. To ensure that enough magnetic strength is provided, each arm of the electromagnet has 1000 wound turns of copper wire 1.2 mm in diameter. A Hall effect Gauss meter, (Kanetec-TM701) with a suitable probe (TM-701PRB), was used to measure the magnetic flux generated by the electromagnet. The MR fluid selected for this study was MRF-132DG manufactured by Lord Corporation. For more details of the devices, one may consult Ref. [23].

5.1 Modal test

In order to perform modal tests, first a laminated composite plate (800 × 800 mm) made of glass fiber has been fabricated. Glass fiber composite is chosen as its magnetic permeability is zero. The plate was then cut into several strips

Figure 4.
Experimental setup.

(250 × 30 × 2.0 mm) by water jet to make beams. We used two similar beam strips as top and bottom faces and make a box beam with a 2.4 mm gap between two strips for MR fluid. Therefore, the total thickness of the MR-laminated beam became 4.4 mm. In order to maintain the uniform gap and hold the fluid between two strips, 2.4 mm thick spacer was glued on the inside face of one strip. Some mechanical properties of the glass fiber layers are given as follows:

Density = 200 g/m^3, $E_1 = E_2 = 15.5$ GPa, $G_{12} = G_{21} = 6.5$ GPa, $\rho = 1650$ kg/m^3.

In this work, the complex shear modulus of the MR fluid as function of magnetic field and driving frequency has been considered from the results of the work done by Naji et al. [10].

To study the effects of magnetic field on modal response of the MR-laminated beams, different magnetic fields, from 0 to 2000 Gauss, have been applied to the specimens.

5.2 Impact tests

To conduct the impact tests, the same electromagnet as described in Section 5.1 has been used. However, for impact tests, the specimens are fabricated as box-aluminum beam with length 400 mm and width 30 mm filled by MR fluid. The thickness of each aluminum layer and MR layer was 1 mm. The density of aluminum was 2700 kg/m^3 and that of the MR fluid was 3500 kg/m^3.

The impact tests were performed by dropping a 5.0 g mass from different heights (0.5, 1.0, and 1.5 m) on the tip of the MR-aluminum cantilever beam. In order to investigate the effect of magnetic field on the impact response of the MR-aluminum beam, the impact tests have been repeated for three levels of magnetic fields, 0, 1000, and 2000 Gauss.

6. Results

In the following two subsections, the analytical results and experimental ones followed by brief explanations are provided.

6.1 Modal tests

The results obtained by analytical approach described in Section 4 and the results extracted from the experimental work are given in **Table 2**. The first three

Magnetic field (Gauss)	Mode	Experimental freq. (Hz)	Analytical results freq. (Hz)
0	1	11.0	11.70
	2	66.5	68.10
	3	185.5	187.36
400	1	11.5	12.03
	2	68.5	69.73
	3	187.0	191.49
800	1	12.0	12.39
	2	70.0	71.96
	3	190.5	192.79
1200	1	11.5	11.76
	2	71.5	73.00
	3	192.0	194.50
1600	1	11.0	11.165
	2	73.0	75.34
	3	193.0	197.83
2000	1	10.0	10.18
	2	75.5	78.67
	3	194.5	198.58

Table 2.
Experimental and analytical natural frequencies of MR beam.

natural frequencies which were extracted from the peak of vibration response spectrum subject to three levels of magnetic fields are compared with analytical ones.

As it is shown, the analytical results provide sufficient agreement with experimental ones for most of the cases.

An important feature that one may conclude from these results is noting that intensifying magnetic field increases the natural frequencies of the MR beams.

Increasing the natural frequencies of MR beams by increasing the magnetic field can be explained by noting that increasing magnetic field increases the stiffness of the MR fluid, which leads to increasing the total stiffness of the MR beam and, in turn, increasing the natural frequencies. However, an exception is shifting the fundamental frequency of the MR beam after 800 Gauss, which is in contradiction with intuitive sense. To interpret this phenomenon, one may note that the loss factor at high magnetic field jumps up dramatically [10], so increasing damping is dominated by increasing stiffness of MR fluid, and higher damping dictates vibration behavior.

6.2 Effects of the MR layer thickness

In order to investigate the effect of MR fluid thickness on the modal response of the MR beams, the first three modes have been computed for different thickness ratios of the MR layer to base material under different applied magnetic fields. The results are given in **Table 3**, where h_1 is the thickness of composite laminated which

Mode	Magnetic field (Gauss)	Natural frequencies of five modes for three different magnetic fields			
		$h_2/h_1 = 1/4$	$h_2/h_1 = 1/2$	$h_2/h_1 = 1$	$h_2/h_1 = 2$
Mode 1	0	9.588	9.432	9.261	9.071
	1000	14.381	12.913	12.758	12.308
	2000	10.826	9.537	9.611	9.934
Mode 2	0	40.349	36.217	33.051	30.624
	1000	44.550	40.574	37.539	34.566
	2000	46.026	43.417	41.308	40.204
Mode 3	0	89.108	77.951	68.818	61.086
	1000	99.501	86.666	76.320	68.195
	2000	101.465	89.037	88.784	81.006

Table 3.
Effect of layer ratio on the natural frequencies of MR-laminated beam.

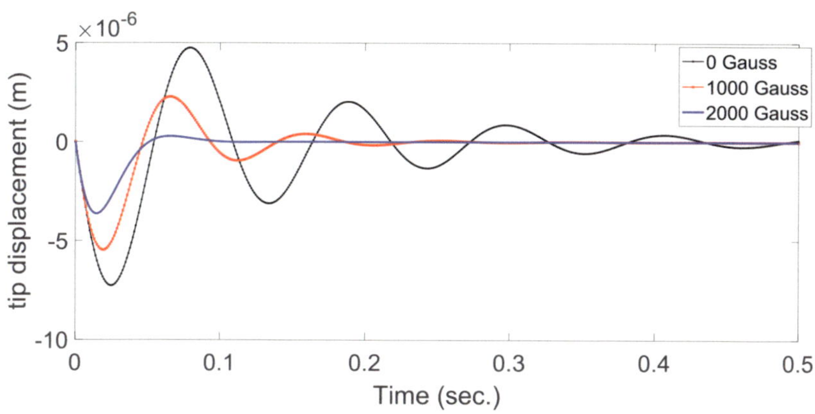

Figure 5.
Impact test for dropping mass from 0.5 m.

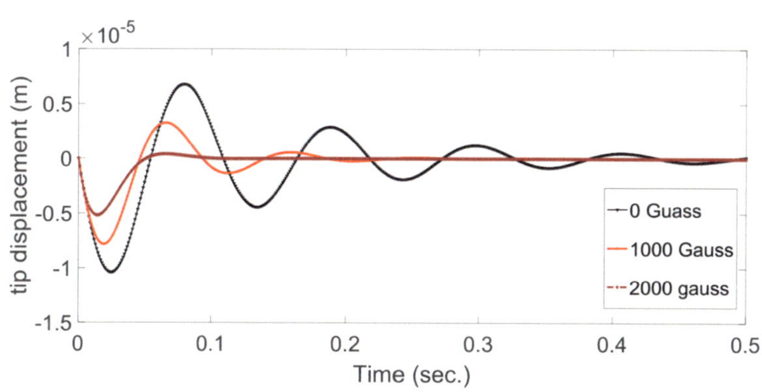

Figure 6.
Impact test for dropping mass from 1.0 m.

is a summation of the upper and lower layers and h_2 is the thickness of MR fluid in middle layer.

The results generally demonstrate that increase in thickness of the MR layer decreases the natural frequencies of all the three modes. This is because in general, the stiffness of MR fluid is lower than the stiffness of the base composite material.

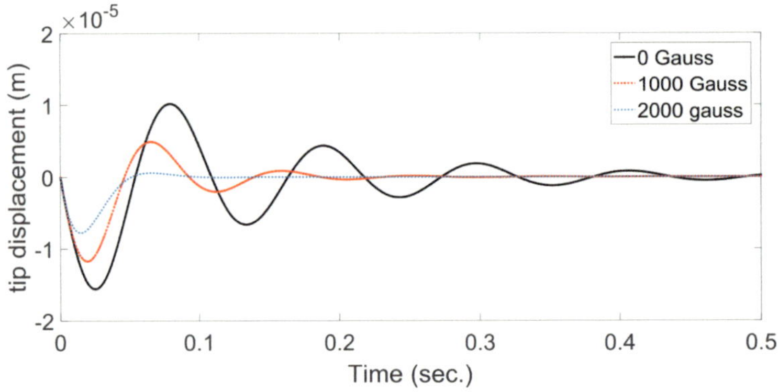

Figure 7.
Impact test for dropping mass from 1.5 m.

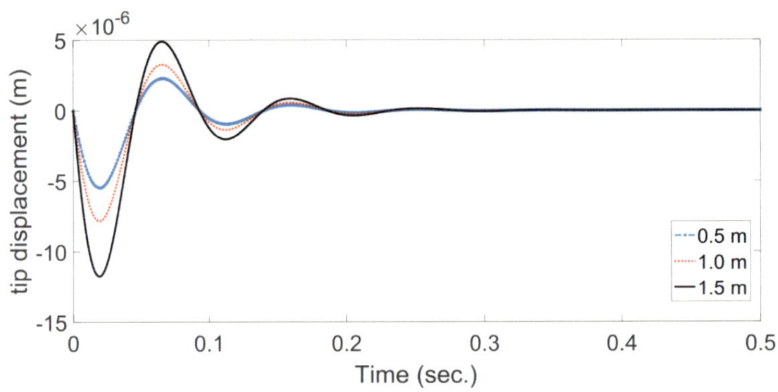

Figure 8.
Impact test for different dropping heights at 1000 Gauss.

6.3 Impact tests

The MR-aluminum beams are subjected to dropping mass as described in Section 5.2. The results for dropping mass from 0.5 m on the tip of the MR beam subject to different magnetic fields are shown in **Figure 5**. As it is observed, increasing the magnetic field makes the MR beam stiffer and thus reduces sharply the settling time.

The impact test results for dropping mass from 1.0 to 1.5 m on the tip of the MR beam for different magnetic fields is shown in **Figures 6** and **7**, where once again it is realized that increasing magnetic field reduces the settling time of the beam response.

To study the effect of magnetic field on the response of the MR beam subject to different impact loads, 1000 Gauss of magnetic field is applied to the MR beam, and its response of different dropping heights is measured and shown in **Figure 8**. As it is observed, increasing the level of impact force increases the amplitude of oscillation.

7. Conclusions

The effect of magnetic field on the vibration and impact responses of the MR beams has been investigated. For modeling purposes, the layerwise displacement theory was employed to overcome some challenges in the modeling of MR beams.

The MR-laminated composite beams in which the top and bottom layers are made of glass-fiber laminated composites and the middle layer, is filled with MR fluids. Experimental tests have been conducted to validate the analytical results and show the performance of the MR-laminated beam for different magnetic fields. It was observed that increasing the magnetic field up to 800 Gauss increases the natural frequencies of the MR-laminated beam. However, beyond the 800 Gauss, the fundamental frequency of the MR-laminated beam begins to drop. The influence of MR layer thickness on the vibration behavior of MR-laminated beam was examined. Adding thickness of the MR layer affected decreases the natural frequencies of the first three modes.

In another study, a three-layered aluminum beam composed of aluminum layers at the top and bottom and the middle layer filled with MR fluid has been investigated for impact loadings. It was realized that increasing the magnetic field reduces the settling time of vibration for MR-aluminum beam. Also, for a constant magnetic field, increasing the level of impact load leads to increasing the amplitude of vibration as it was expected.

Acknowledgements

The authors wish to thank the partial support provided by the Sharif University of Technology, International Campus, on Kish Island.

Author details

Abolghassem Zabihollah*, Jalil Naji and Shahin Zareie
School of Science and Engineering, Sharif University of Technology, International Campus, Kish Island, Iran

*Address all correspondence to: zabihollah@sharif.edu

References

[1] Fattahi SJ, Zabihollah A, Zareie S. Vibration monitoring of wind turbine blade using fiber bragg grating. Wind Engineering. Dec 2010;**34**(6):721-731

[2] Zareie S, Zabihollah A, Azizi A. Buckling control of morphing composite airfoil structure using multi-stable laminate by piezoelectric sensors/actuators. Proceedings of SPIE-The International Society for Optical Engineering. 2011;**7978**

[3] Zareie S, Zabihollah A, Azizi A. Buckling control of morphing composite airfoil structure using multi-stable laminate by piezoelectric sensors/actuators. In: SPIE Smart Structures and Materials + Nondestructive Evaluation and Health Monitoring. 2011

[4] Pol MH, Zabihollah A, Zareie S, Liaghat G. Effects of nano-particles concentration on dynamic response of laminated nanocomposite beam/nano daleliu koncentracijos itaka dinaminei nanokompozicinio laminuoto strypo reakcijai. Mechanika. 2013;**19**(1):53-58

[5] Zabihollah A, Zareie S, Latifi-Navid M, Ghaffari H. Effects of ply-drop off on forced vibraion response of non-uniform thickness laminated composite beams. In: IEEE Toronto International Conference. Vol. 71. 2009. pp. 176-181

[6] Pol MH, Zabihollah A, Zareie S, Liaghat G. On dynamic response of laminated nanocomposite beam/Nano daleliu koncentracijos itaka dinaminei nanokompozicinio laminuoto strypo reakcijai. Mechanika. 2013;**19**(1):53-58

[7] Zabihollah A, Zareie S. Optimal design of adaptive laminated beam using layerwise finite element. Journal of Sensors. 2011;**2011**

[8] Zabihollah A. Analysis, Design Optimization and Vibration Suppression of Smart Laminated Beams. Concordia University; 2007

[9] Powell LA, Hu W, Wereley NM. Magnetorheological fluid composites synthesized for helicopter landing gear applications. Journal of Intelligent Material Systems and Structures. Jun 2013;**24**(9):1043-1048

[10] Sung K-G, Choi S-B, Lee H-G, Min K-W, Lee S-H. Performance comparison of MR dampers with three different working modes: Shear, flow and mixed mode. International Journal of Modern Physics B. 2005;**19**(07n09):1556-1562

[11] Naji J, Zabihollah A, Behzad M. Vibration characteristics of laminated composite beams with magnetorheological layer using layerwise theory. Mechanics of Advanced Materials and Structures. 2017:1-10

[12] Skalski Paweland Kalita K. Role of magnetorheological fluids and elastomers in Today's world. Acta Mechanica et Automatica. 2017;**11**(4):267-274

[13] Cheng M, Chen Z. Semi-active helicopter ground resonance suppression using magnetorheological technology. In: ASME 2017 Conference on Smart Materials, Adaptive Structures and Intelligent Systems. 2017. p. V002T03A015

[14] Azizi A, Durali L, Rad FP, Zareie S. Control of vibration suppression of a smart beam by pizoelectric elements. In: 2nd International Conference on Environmental and Computer Science, ICECS. 2009. p. 2009

[15] Yalcintas M, Dai H. Vibration suppression capabilities of magnetorheological materials based adaptive structures. Smart Materials and Structures. 2003;**13**(1):1-11

[16] Sapiński B, Snamina J. Vibration control capabilities of a cantilever beam with a magnetorheological fluid. Mechanics. 2008;**27**:70-75

[17] Sapiński B, Snamina J, Romaszko M. The influence of a magnetic field on vibration parameters of a cantilever sandwich beam with embedded MR fluid. Acta Mechanica et Automatica. 2012;**6**(1):53-56

[18] Sapiński B, Horak W, Szczęch M. Investigation of MR fluids in the oscillatory squeeze mode. Acta Mechanica. 2013;**7**(2):111-116

[19] Rajamohan V, Sedaghati R, Rakheja S. Vibration analysis of a multi-layer beam containing magnetorheological fluid. Smart Materials and Structures. 2009;**19**(1):15013

[20] Ramamoorthy M, Rajamohan V, AK J. Vibration analysis of a partially treated laminated composite magnetorheological fluid sandwich plate. Journal of Vibration and Control. 2016;**22**(3):869-895

[21] Payganeh G, Malekzadeh K, Malek-Mohammadi H. Free vibration of sandwich panels with smart magneto-rheological layers and flexible cores. Journal of Solid Mechanics. 2016;**8**(1):12-30

[22] Aguib S, Nour A, Djedid T, Bossis G, Chikh N. Forced transverse vibration of composite sandwich beam with magnetorheological elastomer core. Journal of Mechanical Science and Technology. 2016;**30**(1):15-24

[23] Naji J, Zabihollah A, Behzad M. Vibration behavior of laminated composite beams integrated with magnetorheological fluid layer. Journal of Mechanics. 2017;**33**(4):417-425

[24] Naji J, Zabihollah A, Behzad M. Vibration characteristics of laminated composite beams with magnetorheological layer using layerwise theory. Mechanics of Advanced Materials and Structures. 2018;**25**(3):202-211

[25] Naji J, Zabihollah A, Behzad M. Layerwise theory in modeling of magnetorheological laminated beams and identification of magnetorheological fluid. Mechanics Research Communications. 2016;**77**:50-59

[26] Momeni S, Zabihollah A, Behzad M. Development of an accurate finite element model for N-layer MR-laminated beams using a layerwise theory. Mechanics of Advanced Materials and Structures. 2018;**25**(13):1148-1155

[27] Momeni S, Zabihollah A, Behzad M. A finite element model for tapered laminated beams incorporated with magnetorheological fluid using a layerwise model under random excitations. Mechanics of Advanced Materials and Structures. 2018:1-8

[28] Reddy JN. Mechanics of Laminated Plates: Theory and Analysis. Boca Raton, FL: CRC Press; 1997

[29] Iqbal MF. Application of Magneto-Rheological Dampers to Control Dynamic Response of Buildings. Concordia University; 2009

[30] Chu SY, Soong TT, Reinhorn AM. Active, Hybrid and Semi-Active Structural Control—A Design and Implementation Handbook. John Wiley & Sons; 2005. ISBN: 978-0-470-01352-6

[31] Bardzokas DI, Filshtinsky ML, Filshtinsky LA. Mathematical Methods in Electro-Magneto-Elasticity. Springer: Science & Business Media; 2007. Vol. 32. pp. 373-388

[32] Ashour ON, Kinder D, Giurgiutiu V, Rogers CA. Manufacturing and characterization of magnetorheological

fluids. Smart Materials and Structures. 1997;**97**:174-184

[33] Lee DC, Smith DK, Heitsch AT, Korgel BA. Colloidal magnetic nanocrystals: Synthesis, properties and applications. Annual Reports Section "C" (Physical Chemistry). 2007;**103**:351-402

[34] Carlson JD. MR fluids and devices in the real world. International Journal of Modern Physics B. 2005;**19**(07n09):1463-1470

[35] Ashour O, Rogers CA, Kordonsky W. Magnetorheological fluids: Materials, characterization, and devices. Journal of Intelligent Material Systems and Structures. 1996;7(2):123-130

Chapter 6

Applications of Artificial Intelligence Techniques in Optimizing Drilling

Mohammadreza Koopialipoor and Amin Noorbakhsh

Abstract

Artificial intelligence has transformed the industrial operations. One of the important applications of artificial intelligence is reducing the computational costs of optimization. Various algorithms based on their assumptions to solve problems have been presented and investigated, each of which having assumptions to solve the problems. In this chapter, firstly, the concept of optimization is fully explained. Then, an artificial bee colony (ABC) algorithm is used on a case study in the drilling industry. This algorithm optimizes the problem of study in combination with ANN modeling. At the end, various models are fully developed and discussed. The results of the algorithm show that by better understanding the drilling data, the conditions can be improved.

Keywords: optimization, ROP, ABC algorithm, prediction, ANN

1. Introduction

Optimization is the process of setting decision variable values in such a way that the objective in question is optimized. The optimal solution is a set of decision variables that maximizes or minimizes the objective function while satisfying the constraints. In general, optimal solution is obtained when the corresponding values of the decision variables yield the best value of the objective function, while satisfying all the model constraints.

Apart from the gradient-based optimization methods, some new optimization methods have also been proposed that help solve complex problems. In the available classifications, these methods are recognized as "intelligent optimization," "optimization and evolutionary computing," or "intelligent search." One of the advantages of these algorithms is that they can find the optimal point without any need to use objective function derivatives. Moreover, compared to the gradient-based methods, they are less likely to be trapped in local optima.

Optimization algorithms are classified into two types: exact algorithms and approximate algorithms. Exact algorithms are capable of precisely finding optimal solutions, but they are not applicable for complicated optimization problems, and their solution time increases exponentially in such problems. Approximate algorithms can find close-to-optimal solutions for difficult optimization problems within a short period of time [1].

There are two types of approximate algorithms: heuristics and metaheuristics. Two main shortcomings of the heuristic algorithms are (1) high possibility of being trapped into local optima and (2) performance degradation in practical applications on complex problems. Metaheuristic algorithms are introduced to eliminate the problems associated with heuristic algorithms. In fact, metaheuristic algorithms are approximate optimization algorithms that enjoy specific mechanisms to exit local optima and can be applied on an extensive range of optimization problems.

2. Methodology

2.1 Optimization model

The decision-making process consists of three steps: problem formulation, problem modeling, and problem optimization. A variety of optimization models are actually applied to formulate and solve decision-making problems (**Figure 1**). The most successful models used in this regard include mathematical programming and constraint programming models.

2.2 Optimization method

The optimization methods are presented in **Figure 2**. Since the problem is complicated, exact or approximate methods are used to solve it. The exact methods provide optimal solutions and guarantee optimality. Approximate methods lead to favorable and near-optimal solutions, but they do not guarantee optimality.

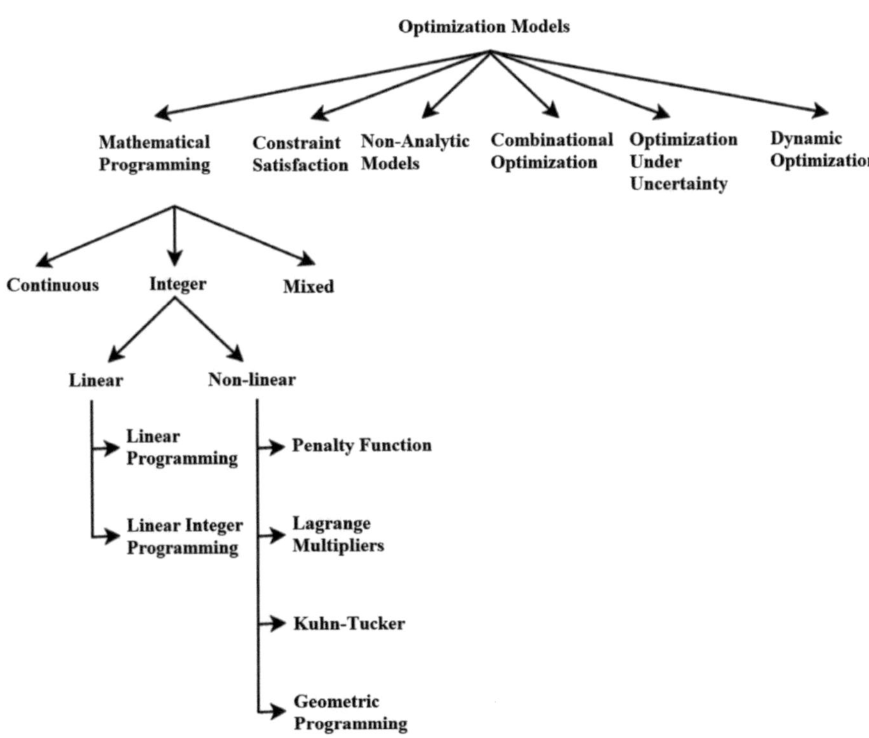

Figure 1.
Optimization models.

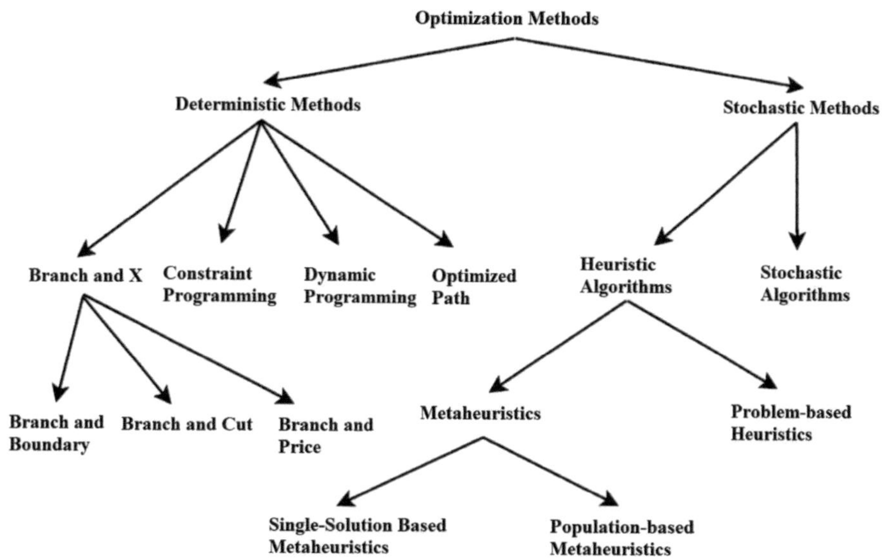

Figure 2.
Optimization methods.

3. Theoretical foundations

3.1 Theoretical foundations of optimization

Any problem in the real world has the potential to be formulated as an optimization problem. Generally, all optimization problems with explicit objectives can be expressed as nonlinearly constrained optimization problem as presented in Eq. (1).

$$\begin{aligned} &\max \ or \ \min f(x), \qquad x = (x_1, x_2, \dots, x_n)^T \in \mathbb{R}^n \\ &Subject \ to \ \phi_j(x) = 0, \quad j = 1, 2, \dots, M \\ &\psi_k(x) \leq 0, \ k = 1, 2, \dots, N \end{aligned} \tag{1}$$

where $f(x)$, $\phi_j(x)$, and $\psi_k(x)$ are the scalar functions of the column vector x. The x_i elements of the vector x are the design variables, or the decision variables, that could be either continuous, discrete, or mixed of the two. The vector x is often referred to as the decision vector, which varies in an n-dimensional space $\mathbb{R}^n$. The function f (x) is called the objective function or the energy function. The objective function is called the cost function in minimization problems and fitness function in maximization problems. Moreover, $\phi_j(x)$ are constraints in terms of M equalities and $\psi_k(x)$ are constraints in terms of N inequalities. Thus, in general, we will have a total of M + N constraints. The space spanned by the decision variables is known as the search space, and the space spanned by the objective function value is called the solution space. The optimization problem maps the search space on the solution space.

3.1.1 Norms

For a vector v, p-norm is denoted by $\|v\|_p$ and defined as Eq. (2).

$$\|v\|_p = \left(\sum_{i=1}^{n} |v_i|^p \right)^{1/p} \tag{2}$$

where p is a positive integer. According to this definition, one can understand that a p-norm satisfies the following conditions: $\| \|v\| \geq 0$ for all $\| \|v\| = 0$ if and only if v = 0. This shows the nonnegativeness of p-norm. In addition, for each real number α, we have the scaling condition $\|\alpha v\| = \alpha\|v\|$. Three most commonly used norms are 1-, 2-, and infinity norms, when p is equal to 1, 2, and ∞, respectively.

3.1.2 Eigenvalues and eigenvectors

The eigenvectors for a square matrix $[A]_{n\times n}$ are defined as Eq. (3).

$$(A - \lambda I)u = 0 \tag{3}$$

where I is a unitary matrix with the same size as A. All the nontrivial solutions are obtained from Eq. (4).

$$\begin{bmatrix} a_{11} - \lambda & a_{12} & \cdots & a_{1n} \\ a_{21} & a_{22} - \lambda & & a_{2n} \\ \vdots & \vdots & \ddots & \vdots \\ a_{n1} & a_{n2} & \cdots & a_{nn} - \lambda \end{bmatrix} = 0 \quad \det|A - \lambda I| = 0 \tag{4}$$

which can be written as a polynomial in form of Eq. (4).

$$\lambda^n + \alpha_{n-1}\lambda^{n-1} + \ldots + \alpha_1\lambda + \alpha_0 = (\lambda - \lambda_1)(\lambda - \lambda_2)\ldots(\lambda - \lambda_n) = 0 \tag{5}$$

where λ_i are eigenvalues and could be complex numbers as well. For each eigenvalue λi, we have a corresponding eigenvector u_i, whose direction can be defined uniquely, but the eigenvector length will not be unique, since any nonzero multiple of vector u can satisfy Eq. (3) and can thus be considered as an eigenvector.

3.1.3 Spectral radius of the matrix

The spectral radius of a square matrix is another important concepts associated with eigenvalues of matrices. Assuming that λ_i are eigenvalues of the square matrix A, the spectral radius of the matrix $\rho(A)$ will be defined as Eq. (6).

$$\rho(A) = \max\{|\lambda_i|\} \tag{6}$$

which is equal to the maximum absolute value of all eigenvectors. Geometrically speaking, if we draw all the eigenvalues of matrix A on a complex plane and then draw a circle on the plane, in such a way that it encloses all the eigenvalues, then the minimum radius of such a circle is referred to as the spectral radius. Spectral radius is useful in determining the stability or instability of iterative algorithms.

3.1.4 Hessian matrix

The gradient vector of a multivariate function f (x) is defined according to Eq. (7),

$$G_1(x) \equiv \nabla f(x) \equiv (\partial f/\partial x_1, \partial f/\partial x_2, \ldots, \partial f/\partial x_n)^T \tag{7}$$

where $x = (x_1, x_2, \ldots, x_n)$ is a vector. Since $\nabla f(x)$is a linear function, it is defined as the vector constant k, and the linear function is generated from Eq. (8).

$$f(x) = k^T x + b \tag{8}$$

where b is a vector constant.

Second derivative of a general function f (x) of a matrix n × n is called the Hessian matrix,

$$G_2(x) = \nabla^2 f(x) = \begin{pmatrix} \partial^2 f/\partial x_1^2 & \cdots & \partial^2 f/\partial x_1 \partial x_n \\ \vdots & \ddots & \vdots \\ \partial^2 f/\partial x_n \partial x_1 & \cdots & \partial^2 f/\partial x_n^2 \end{pmatrix} \tag{9}$$

3.1.5 Convexity

Linear programming problems are usually classified according to the convexity of their defining functions. Geometrically speaking, an object is called convex when for any two points within the object, every point on the straight line connecting them also lies within the object (**Figure 3**). Mathematically, a set $S \in \mathbb{R}^n$ within the space of a real vector is called a convex set when Eq. (10) holds true.

$$tx + (1-t)y \in S, \quad \forall (x,y) \in S, t \in [0,1] \tag{10}$$

A function $f(x)$ defined on the convex set Ω is called convex if and only if:

$$\begin{aligned} &f(\alpha x + \beta y) \le f(\alpha x) + f(\beta y), \quad \forall (x,y) \in \Omega \\ &\alpha \ge 0, \beta \ge 0, \alpha + \beta = 1 \end{aligned} \tag{11}$$

An interesting feature of the convex function f is that it ensures that the gradient at a point $\left.\frac{df}{dx}\right|_{x^*} = o$ approaches zero. In this case, x^* is an absolute minimum point for f.

3.1.6 Optimality criteria

Mathematical programming includes several concepts. Here, we will first introduce three related concepts: feasible solution, strong local maximum, and weak local maximum.

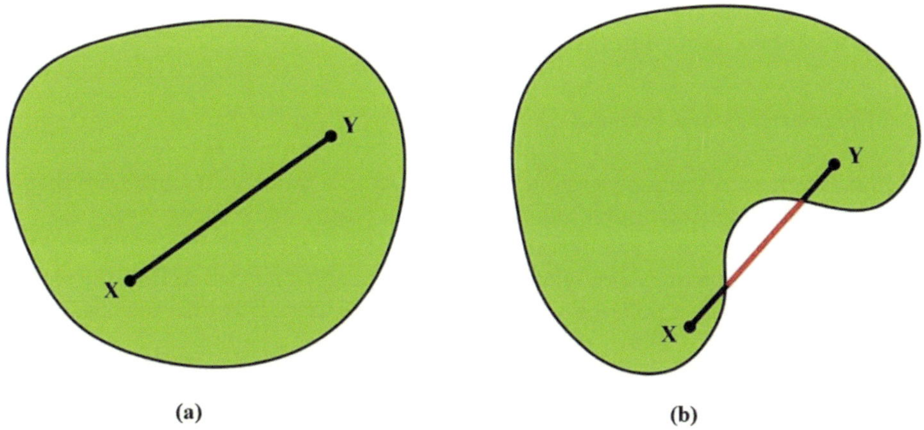

Figure 3.
Convex object (a) and nonconvex object (b).

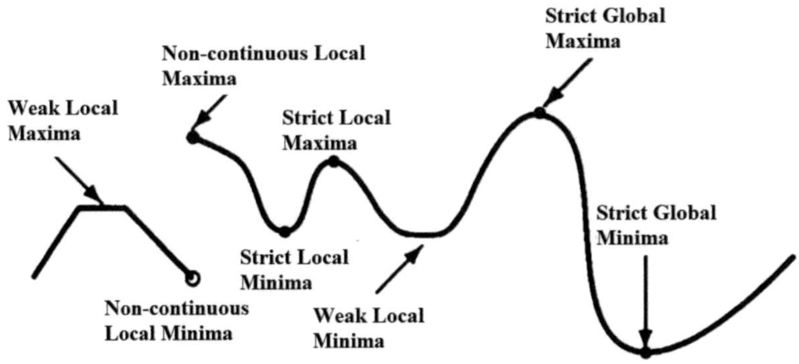

Figure 4.
Strong and weak local minima and maxima.

Point X that satisfies all the constraints of the problem is called a feasible solution. The set of all feasible points will form the feasible region.

Point x is a strong local maximum if f (x) is defined in δ neighborhood $N(\delta, x_*)$and satisfies$f(x_*)>f(x)$ for each $\forall u \in N(\delta, x_*)$ where $\forall u \in N(\delta, x_*)$ and $u \neq x_*$.The inclusion of equality in the condition$f(x_*) \geq f(x)$ will define x as a weak local maximum. A schematic view of strong and weak local maxima and minima is presented in **Figure 4**.

3.1.7 Computational complexity

The efficiency of an algorithm is usually measured by algorithmic complexities or computational complexities. Such complexities are often referred to as Kolmogorov complexity in literature. For a given problem with complexity of n, this complexity is represented by big-O notations, for example, $O(n^2)$or $O(n \log n)$ [1]. For two functions f (x) and g (x), if we have,

$$\lim_{x \to x_0} f(x)/g(x) \to K; \qquad f = O(g) \tag{12}$$

where K is a finite and nonzero value. The big-O notation indicates that f is asymptotically equivalent to the order of g. If the limit value is K = 1, it can be argued that f is of the same order as g [1]. The small-o notation is applied when the limit tends to be zero,

$$\lim_{x \to x_0} f(x)/g(\mathrm{x}) \to 0; \qquad f = o(g) \tag{13}$$

3.1.8 Nondeterministic polynomial (NP) problems

In mathematical programming, an easy or tractable problem is a problem that can be solved using a computer algorithm, with a reasonable solution time, as a polynomial function of problem size n. An algorithm is referred to as a P-problem, or a polynomial-time problem, when the number of steps needed to find the solution is represented by a polynomial in terms of n and there is at least one algorithm to solve it.

On the other hand, a hard or intractable problem is a problem whose solution time is an exponential function of n. In case the solution to a polynomial problem is estimated in polynomial time, then it is called a nondeterministic polynomial. But it should be noted that there is no specific rule for making such a guess. As a result, the estimated solutions cannot be guaranteed to be optimal or even near-optimal solutions. In fact, there is no specific algorithm for solving hard-NP problems, and only approximate or heuristic solutions are applicable. Therefore, heuristic and metaheuristic methods can provide us the near-optimal/suboptimal responses with acceptable accuracy.

A given problem can be called NP-complete if it is actually an NP-hard problem, and other NP problems can be reduced to it using certain reduction algorithms. The reduction algorithm has a polynomial time. The traveling salesman problem can be counted as an example of NP-hard problem, which aims to find the shortest route or the lowest traveling cost to visit all n cities once and then return to the starting city.

3.2 Theoretical foundations of metaheuristic optimization

Two opposite criteria should be taken into account in development of a metaheuristic algorithm: (1) exploration of the search space and (2) exploitation of the best solution (**Figure 5**).

Promising areas are specified by good solutions obtained. In intensification, the promising regions are explored accurately to find better solutions. In diversification, attempts are made to make sure that all regions of the search space are explored.

In the exploration approach, random algorithms are the best algorithms for searching. Random algorithms generate a random solution in each iteration and completely exploit the search space in this way.

3.2.1 Representation

The simulation of any metaheuristic algorithm requires an encoding method. In other words, the problem statement procedure is referred to as representation. Encoding plays a major role in the productivity and efficiency of any metaheuristic algorithm and is recognized as a necessary step in the algorithm. Additionally, the representation efficiency depends on the search operators (neighborhood, recombination, etc.). In fact, when defining a representation, we first need to remember how the problem is evaluated and how the search operator will work. A representation needs to have the following characteristics:

Completeness: It is one of the main characteristics of representation; in the sense that all the solutions of a given problem need to be represented.

Connectivity: It means that a search path must exist between any two solutions in the search space.

Efficiency: Representation should be easily available to search operators.

Representations can be divided into two types in terms of their structure: linear and nonlinear. In this study, linear representation has been used. Some linear representations include the following:

Binary encoding: It is performed using binary alphabets.

Continuous encoding: In continuous optimization problems, encoding is performed based on real numbers.

Discrete encoding: It is used for discrete optimization problems such as the assignment problem.

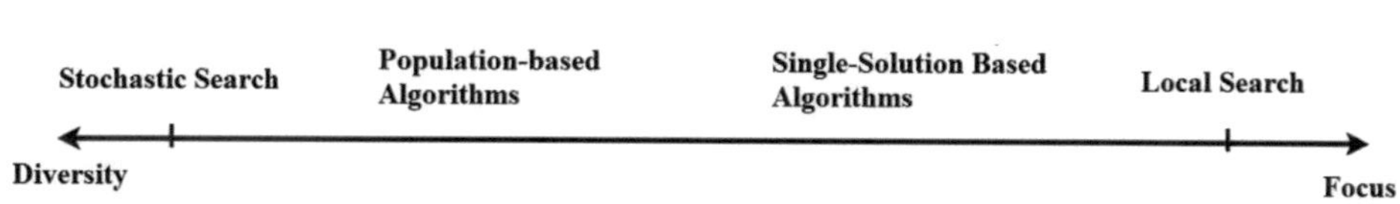

Figure 5.
Metaheuristic algorithm design space.

Permutation encoding: It is used in problems where the objective is to find a permutation.

Random Key: This type of encoding converts real numbers into a permutation.

Diploid representation: In the diploid representation, two values are considered for each subset of the decision vector.

3.2.2 Objective function

The objective function generates a real number for any solution in the search space. This number describes the quality or the fitness of the solution. The objective function is an important element in development of a metaheuristic algorithm that directs the search toward the best solution. If the objective function is wrongly defined, it will generate unacceptable solutions. In the present work, the objective function is maximization of drilling penetration rate.

3.2.3 Constraint

Constraint handling is another critical issue for the efficient design of metaheuristic algorithms. In fact, many continuous or discrete optimization problems are constrained. As mentioned earlier, constraints might be linear or nonlinear, equal or unequal. Constraints can mostly be applied to the decision variables or objective function. Some constraint handling strategies are presented in this section; these strategies can be categorized as follows:

Reject strategy: In this approach, infeasible solutions are rejected, and only the feasible ones are taken into account.

Penalizing strategy: In this strategy, infeasible solutions obtained during the search process are preserved in the search space. This strategy is the most popular strategy used to handle constraints. This strategy uses the penalizing strategy to transform problems with constraints into a problem with no constraint.

Repairing strategy: In this strategy, infeasible solutions turn into feasible solutions.

Preserving strategy: In this strategy, specific operators are used to generate feasible solutions alone.

3.2.4 Search strategy

Search strategy is of particular importance in metaheuristic algorithms. This strategy carries out the search process without using the derivative of the problem. Some of the leading search models are listed below.

Golden Section search: This is a technique used to find the extremum (maximum and minimum) of a unimodal function by narrowing the range of values inside which the extremum is known to exist.

Random search: Random search is a numerical optimization method independent of the gradient and hence can be used for noncontinuous or nondifferentiable functions.

Nelder-Mead method: The Nelder-Mead method, also known as downhill simplex, is usually used for nonlinear optimization. This method is a numerical method that can converge to nonstationary points.

3.2.5 Classification of metaheuristic algorithms

The criteria used for classification of metaheuristic algorithms are as follows:

Nature-inspired vs. nonnature inspiration: Many of the metaheuristic algorithms are inspired by natural processes. Evolutionary algorithms and artificial immune systems, ranging from biological behavior of bee, social behavior of bird flocking, and physical behavior of materials in simulated annealing to human-sociopolitical behavior in imperialist competitive algorithm, belong to these nature-inspired algorithms.

Memory usage versus memoryless methods: Some metaheuristic algorithms are memoryless. These algorithms do not store data dynamically during search time. Simulated annealing lies in this category of algorithms, while some other metaheuristic algorithms use information explored during the search process. Short-term and long-term memory used in tabu search algorithm are of this type.

Deterministic or stochastic: Deterministic metaheuristic algorithms solve optimization problems through deterministic decision-making (such as local search and tabu search). In stochastic metaheuristic algorithms, several stochastic rules are applied to searching. In deterministic algorithms, the initial solution leads to the generation of a final solution similar to the initial one.

Population-based vs. single-point search algorithms: Single-point algorithms (such as simulated annealing) direct and transmit a single solution throughout the search process, while population-based algorithms (such as particle swarm optimization) will involve the whole solution population. Single-point search algorithms apply an exploitive approach; these algorithms have the power to concentrate searching on the local space. Population-based algorithms have exploratory trajectory and allow for more diversified exploration of the search space.

Iterative or greedy approach: In iterative algorithms, the search starts with an initial set of solutions (population), and the solutions vary in each iteration. In greedy algorithms, the search begins with a null solution, and a decision variable is determined at each step until the final solution is obtained. Most metaheuristic algorithms follow an iterative approach.

4. Review of literature

In this section, firstly, a brief explanation of some of the mostly used metaheuristic algorithms is provided. Next, previous works dealing with prediction and optimization of penetration rate performed by various authors are introduced.

4.1 Literature on metaheuristic optimization

The optimization literature changed dramatically with the advent of metaheuristic algorithms in the 1960s. Alan Turing might be the first to use heuristic algorithms. During the Second World War, Alan Turing and Gordon Welchman managed to design the Bambe machine and finally crack the German Enigma machine in 1940. In 1948, he managed to get a patent for his ideas in the field of intelligent machinery, machine learning, neural network, and evolutionary algorithms.

4.1.1 Genetic algorithm

The genetic algorithm that was developed by John Holland et al. during 1960–1970 is a biological evolutionary model inspired by Charles Darwin's natural selection and survival of the fittest. Holland was the first to use crossover, recombination, mutation, and selection in comparative studies and artificial systems [2]. **Figures 6** and 7 indicate the application of crossover and mutation operators.

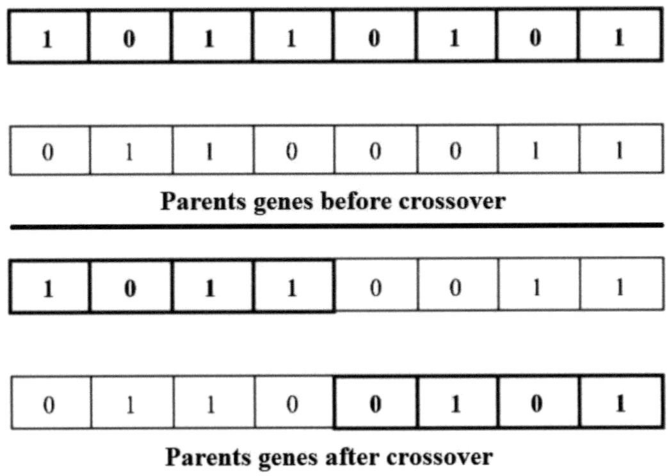

Figure 6.
The schematic view of crossover at a random point [2].

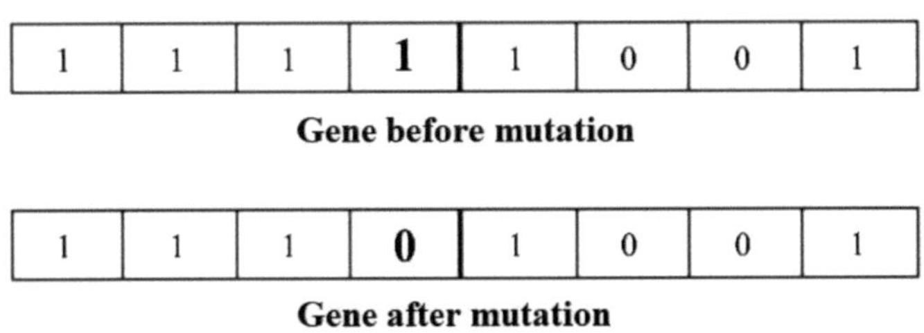

Figure 7.
The schematic view of mutation at a random point [2].

4.1.2 Simulated annealing algorithm

Patrick et al. developed a simulated annealing algorithm to solve optimization problems. When steel is cooled, it develops into a crystallized structure with minimum energy and larger crystalline sizes, and the defects of steel structure are decreased (**Figure 8**) [3].

The search technique used in this algorithm is a movement-based search, which starts from an initial guess at high temperatures and the system cools down with a gradual decrease in temperature. A new movement or solution is accepted if it is better. Otherwise, it will be accepted as a probable solution so that the system can be freed from the local optima trap [3].

4.1.3 Tabu search algorithm

Tabu search was discovered by Glover [4]. It is a memory-based search strategy that searches the memory history as an integrative element. Two important points should be taken into account in this search: (1) how to efficiently use memory and (2) how to integrate the algorithm into other algorithms to develop a superior algorithm. Tabu search is the centralized local search algorithm that uses memory to avoid potential cycles of local solutions to increase search efficiency.

In the algorithm running stages, recent attempts (memory history) are recorded and listed as tabu, such that new solutions should avoid those available in the tabu list. Tabu list is one of the most important concepts in the tabu search method and records the search moves as a recent history, so that any new move must avoid the

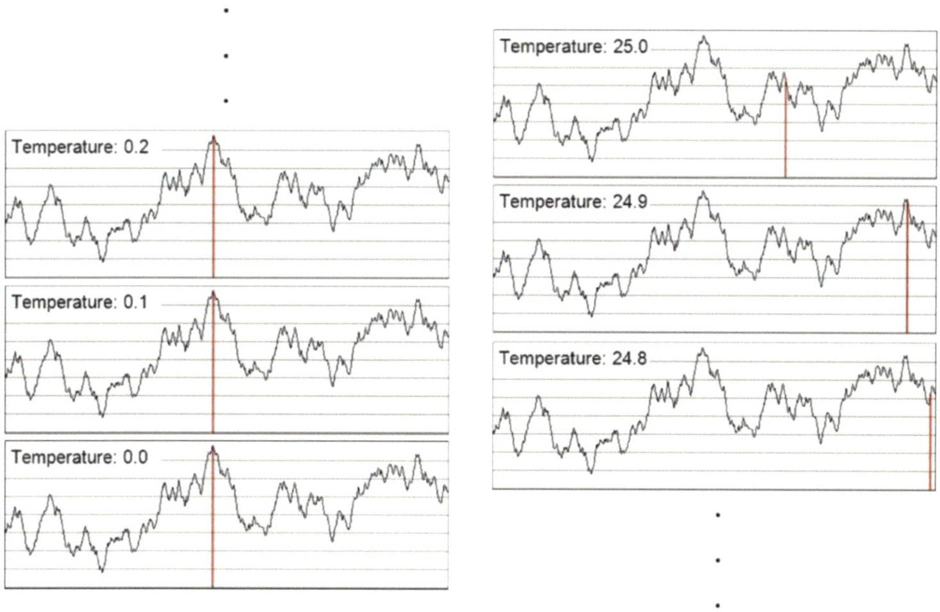

Figure 8.
Simulated annealing search technique.

previous move list. This will also lead to time saving because the previous move is not repeated [4].

4.1.4 Ant colony optimization

When ants find a food source, they use pheromones to mark the food source and the trails to and from it. As more ants cross the same path, that path turns into a preferred path (**Figure 9**). Thus, several preferred paths will emerge during the

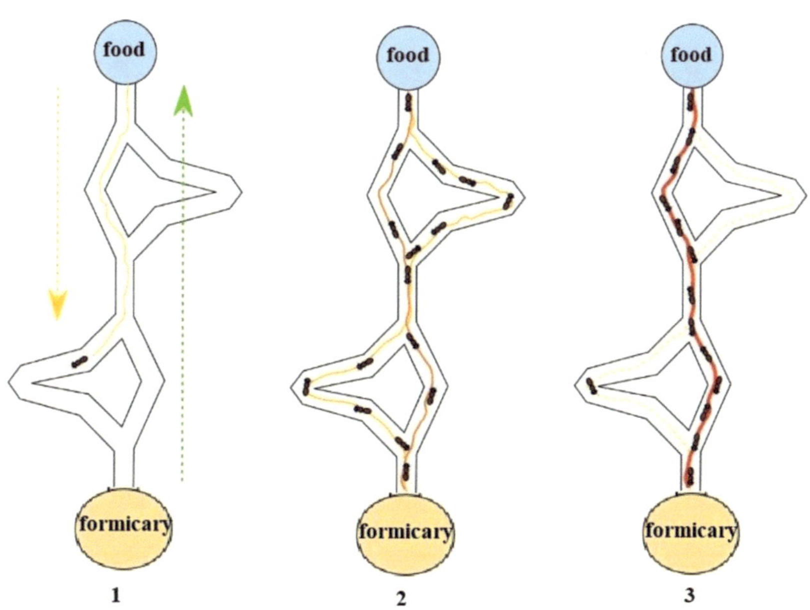

Figure 9.
Ant preferred trail formation process.

process. Using this behavioral property of the ants, scientists have managed to develop a number of robust ant colony optimization methods. Dorigo was known as a pioneer in this field in 1992 [5].

4.1.5 Particle swarm optimization

Sometime later, the particle swarm optimization was developed by [6]. This method is inspired by the collective behavior exhibited by birds, fish, and even humans, which is referred to as swarm intelligence. Particles swarm around the search space based on initial random guess. This swarm communicates the current best and the global best and is updated based on the quality of the solutions. The movement of particles includes two main components: a stochastic component and a deterministic component. A particle is attracted toward the current global best while it has a tendency to move randomly. When a particle finds a location that is better than the previous ones, it updates it as the new best location. **Figure 10** shows the schematic view of the motion of particles [7].

4.1.6 Harmony search

Harmony search was first developed by Geem et al. [8]. Harmony search is a metaheuristic algorithm inspired by music, which is developed based on the observation that the aim of music is to search for a perfect state of harmony. This harmony in music is analogous to find optimality in an optimization process. When a musician wants to play a piece of music, there are three choices:

- Harmony memory accurately plays a piece of famous music on memory.
- Pitch adjusting plays something similar to a famous piece.
- Randomization sets a random or new note [8].

4.1.7 Honeybee algorithm

Honeybee algorithm is another type of optimization algorithm. This algorithm is inspired by the explorative behavior of honeybees, and many variants of this algorithm have already been formulated: honeybee algorithm, virtual bee algorithm, artificial bee colony, and honeybee mating algorithm.

Literature suggests that the honeybee algorithm was first formulated by Sunill Nakrani and Craig Tovey (2004) at Oxford University in order to be used to allocate computers among different clients and web hosting servers [9].

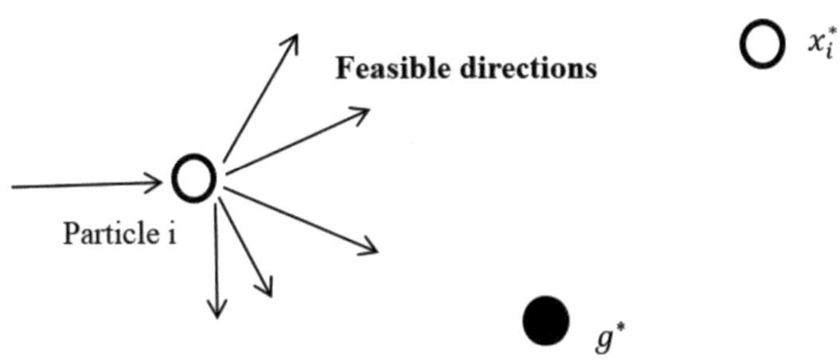

Figure 10.
Schematic representation of particle motion in the particle swarm method.

4.1.8 Big Bang-Big Crunch

Big Bang-Big Crunch was first presented by Erol and Eksin [10]. This approach relies on theories of the evolution of the universe, namely the Big Bang-Big Crunch evolution theory. In the Big Bang phase, energy dissipation causes a state of disorder or chaos, and randomization is known as the principal feature of this stage. In the Big Crunch stage, however, the randomly distributed particles are drawn into an order [10].

4.1.9 Firefly algorithm

The Firefly algorithm was developed by Yang [11] at Cambridge University based on idealization of the flashing characteristics of fireflies. In order to develop the algorithm, the following three idealized rules are used:

All fireflies are unisex, such that a firefly will be attracted to other fireflies, regardless of their gender.

Attractiveness is proportional to its desired brightness, hence for any of the two flashing fireflies, the less brighter firefly will move toward the more brighter one.

The brightness of a firefly can be determined by the landscape of the objective function [11].

4.1.10 Imperialist competitive algorithm

The imperialist competitive algorithm was developed by Atashpaz Gargari and Lucas in 2007. Drawing on mathematical modeling of sociopolitical evolution process, this algorithm provides an approach to solving mathematical optimization problems. During the imperialist competition, weak empires lose their power gradually and are finally eliminated. The imperialist competition makes it possible for us to reach a point where there is only one empire left in the world. This comes to realization when the imperialist competitive algorithm reaches the optimal point of the objective function and stops [12].

4.1.11 Cuckoo search

Cuckoo search is an optimization algorithm developed by Yang and Deb in 2009. This algorithm is inspired by the obligate brood parasitism of some cuckoo species by laying their eggs in the nests of other host birds. The following idealized rules are used for more simplicity:

Each cuckoo lays egg once at a time and puts it in a randomly selected nest.

The best nests with high-quality eggs will carry over to the next generation.

The number of hosts nests is fixed, and the egg laid by a cuckoo can be discovered by the host bird with a probability $P\,a \in (0,1)$. In this case, the host bird will either dump the eggs or just leaves the nest to build a new one somewhere else [13].

4.1.12 Bat algorithm

The bat algorithm is a metaheuristic optimization algorithm developed by Yang [14]. This algorithm is based on the echolocation behavior of microbats with varying pulse rates of emission and loudness. Echolocation is a biological sound tracking system that is used by bats and some other animals, such as dolphins. By idealization of some of the echolocation features, one can develop various bat-inspired algorithms:

All bats use echolocation to sense distance, and they also "know" the difference between food/prey and background barriers in some magical way.

Bats fly randomly with velocity v_i at position x_i with a fixed frequency f_{min}, varying wavelength λ and loudness A_0 to search for prey. They can automatically adjust the wavelength (or frequency) of their emitted pulses and adjust the rate of pulse emission r ∈ [0, 1], depending on the proximity of their target.

Although loudness may vary in many ways, it is assumed that loudness variations range from a large (positive) A_0 to a minimum constant value A_{min} [14].

4.1.13 Charged system search

Charged system search was presented by Kaveh and Talatahari [15] for optimization of mathematical model. Each search agent is referred to as a charged particle, which behaves like a charged sphere with a known radius and a charge proportional to the quality of the produced solution. Thus, the particles are able to exert force on one another and cause other particles move. In addition, exploitation of particle's previous velocity as a consideration of the particle's past performance can be effective in changing the particle position. Newtonian mechanic rules were used to precisely determine these changes from the rules used here provided some sort of balance between the algorithm power at the conclusion and search stage [15].

4.1.14 Krill herd algorithm

The krill herd algorithm was proposed by Gandomi and Alavi [16] to optimize the mathematical model. This algorithm is classified as a swarm intelligence algorithm. This algorithm is inspired by the herding behavior of krill swarms in the process of food finding. In the krill herd algorithm, minimum distance of the krill individual from food and from the highest density of the herd is considered as the objective functions for krill movement. The specific location of the individual krill varies with time depending on the following three actions: movement induced by other krill individuals; foraging activity; and random diffusion (**Figure 11**).

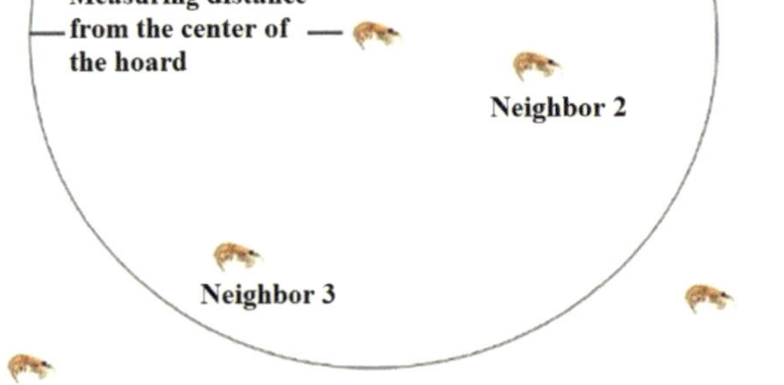

Figure 11.
The effect of neighbors and herd center on the movement of krill [16].

4.1.15 Dolphin echolocation

Dolphin echolocation was first proposed by Kaveh and Farhoudi as a new optimization method. Scientists believe that dolphins are ranked second (after humans) in terms of smartness and intelligence. This optimization method was developed according to echolocation ability of dolphins [17].

4.2 Literature on drilling operations

Drilling operations lead to significant costs during the development of oil and gas fields. Therefore, drilling optimization can decrease the costs of a project and hence increase the profit earned from the oil and gas production. In most of the studies, rate of penetration (ROP) has been considered as the objective function of the optimization process. ROP depends on many factors including well depth, formation characteristics, mud properties, rotational speed of the drill string, etc. Several studies have been conducted to gain a profound insight into the effective parameters on ROP. Maurer [18] introduced an equation for ROP, in which it was accounted for rock cratering mechanisms of roller-cone bits. Galle and Woods [19] proposed a mathematical model for estimating ROP, where formation type, weight on bit, rotational speed of bit, and bit tooth wear were taken as input parameters. Mechem and Fullerton [20] proposed a model with input variables of formation drilling ability, well depth, weight on bit, bit rotational speed, mud pressure, and drilling hydraulics. Bourgoyne and Young [21] used multiple regression analysis to develop an analytical model and also investigated the effects of depth, strength, and compaction of the formation, bit diameter, weight on bit, rotational speed of bit, bit wear, and hydraulic interactions associated with drilling. Bourgoyne and Young [21] introduced a technic for selection of optimum values for weight on bit, rotational speed, bit hydraulics, and calculation of formation pressure through multiple regression analysis of drilling data. Tanseu [22] developed a new method of ROP and bit life optimization based on the interaction of raw data, regression, and an optimization method, using the parameters of bit rotational speed, weight on bit, and hydraulic horsepower. Al-Betairi et al. [23] used multiple regression analysis for optimization of ROP as a function of controllable and uncontrollable variables. They also studied the correlation coefficients and multicollinearity sensitivity of the drilling parameters. Maidla and Ohara [24] introduced a computer software for optimum selection of roller-cone bit type, bit rotational speed, weight on bit, and bit wearing for minimizing drilling costs. Hemphill and Clark [25] studied the effect of mud chemistry on ROP through tests conducted with different types of PDC bits and drilling muds. Fear [26] conducted a series of studies using geological and mud logging data and bit properties in order to develop a correlation for estimating ROP. Ritto et al. [27] introduced a new approach for optimization of ROP as a function of rotational speed at the top and the initial reaction force at the bit, vibration, stress, and fatigue limit of the dynamical system. Alum and Egbon [28] conducted a series of studies, which led to the conclusion that pressure loss in the annulus is the only parameter that affects ROP significantly, and finally, they proposed an analytical model for estimation of ROP based on the model introduced by Bourgoyne and Young. Ping et al. [29] utilized shuffled frog leaping algorithm to optimize ROP as a function of bit rotational velocity, weight on bit, and flow rate. Hankins et al. [30] optimized drilling process of already drilled wells with variables of weight on bit, rotational velocity, bit properties, and hydraulics to minimize drilling costs. Shishavan et al. [31] studied a preliminary managed pressure case to minimize the associated risk and decrease the drilling costs. Wang and Salehi [32] used artificial intelligence for prediction of optimum mud hydraulics during drilling operations and performed sensitivity analysis using forward regression. A variety of artificial intelligence works have recently been conducted in civil and oil engineering [33–36].

In the following sections, a new approach was used for prediction and optimization of ROP, based on artificial neural network (ANN). According to the authors' knowledge, ANN application on ROP optimization has not been widely used by previous studies. The variables used in this study were well depth (D), weight on bit (WOB), bit rotational velocity (N), the ratio of yield point to plastic viscosity (Y_p/PV), and the ratio of 10 min gel strength to 10 s gel strength (10MGS/10SGS). Using ANN technic, several models were developed for prediction of ROP, and the best one was selected according to their performances. Then, an artificial bee colony (ABC) algorithm was used for optimization of ROP based on the selected ANN predictive model, and the drilling parameters were evaluated to determine their effects on ROP.

5. Methodology of the problem of the case study

In the present work, it is aimed to apply neural networks in combination with artificial bee colony (ABC) algorithm on a real case of penetration rate prediction and optimization. The basic definitions regarding the problem of study are provided in the nest subsections. Then, the case used in our work is explained. At the end, ABC algorithm used in the optimization process is described.

5.1 Hydrocarbon reservoir

Hydrocarbon is the general term used for any substance, which is composed of hydrogen and carbon. From clothing to energy, there are different areas in which hydrocarbons serve as the main material. Hydrocarbons are usually extracted from reservoirs located deep in the formation of the earth's crust. Underground hydrocarbon reservoirs, which are also known as oil and gas reservoirs, have been exploited since more than one and half a century ago. And there have been several developments in technologies associated with oil and gas industry [37, 38].

The term hydrocarbon reservoir is used for a large volume of rock containing hydrocarbon either in oil or gas form, which is usually found in deep formation in the earth. This type of reservoir is far different from what most of people imagine when they think about. A hydrocarbon reservoir is not a tank or something like that. In fact, it is a rock having numerous pores, which make it capable of storing fluid. There are two types of hydrocarbon reservoirs: conventional and unconventional [39].

A conventional reservoir consists of porous and permeable rock, which is bounded by an impermeable rock, usually called cap rock. Due to the high pressure in the deep layers, the fluid in the reservoir rock tends to move out of the rock toward lower depths, which usually have lower pressures. The role of cap rock is to seal the rock in order to prevent the hydrocarbon from migrating to low-pressure depths.

Conventional reservoirs were the only type of exploited hydrocarbon reservoirs until the recent years. As the conventional reserves became rare and depleted, oil and gas industries started to study the feasibility of production from unconventional reservoirs. Thanks to the recent developments in the related technologies, production of hydrocarbon from unconventional reservoirs has been started in different locations of the earth. The major difference between conventional and unconventional reservoirs is that in unconventional reservoirs, there is no traditional placement of reservoir and cap rock. The reservoir rock has high κporosity, but because of low permeability, the fluid cannot move out of it and is entrapped into the rock. Since the example of the present work deals with a conventional reservoir, we avoid discussing more about unconventional reservoirs.

In order to produce oil and gas from a reservoir, at the first step, it is required to find a location in which hydrocarbon is accumulated in such a large volume that it can be exploited in an economic way. This exploration step is typically done using seismic technics. In the next step, the location with high probability of having hydrocarbon storage is drilled. The drilled well is called exploration well, and if it reaches a relatively large amount of hydrocarbon, more wells are drilled after preparing a field development plan. The production of the reservoir continues until the production rate falls below an economic criterion, which is usually defined as net present value.

Due to the high pressure of the reservoir rock, the hydrocarbon tends to move toward a lower pressurized region. In order to exploit the entrapped hydrocarbon and providing a flow path, one or more wells are needed. The well is drilled deep into the rocks, and after passing the cap rock, it reaches the reservoir rock. Then, due to the pressure difference between the rock and surface, the hydrocarbons start to move from the reservoir to the surface through the drilled well. Sometimes the pressure difference is not so large that the fluid can reach the surface. In these cases, some technics, called artificial lift methods, are used to increase the energy for delivering the fluid to higher altitude. After extraction of hydrocarbon, it is delivered to treatment facilities and the next steps are designed according to the producer company's plan.

5.2 Drilling operations

As mentioned above, exploitation of oil and gas reservoirs typically consists of the three types of operation: exploration, drilling, and production. The drilling phase involves costly operations, which consume a high portion of the capital expenditure of the field development. Therefore, optimizing the operations associated with drilling can reduce the investments significantly, increasing the net present value of the project [40].

In the early years of oil and gas industry, the wells were drilled using percussion table tools. These technics became inefficient as demand for drilling deep and hence more pressurized formations increased. In the early twentieth century, rotary drilling technic was introduced to oil and gas industries and it paved the way for drilling faster and deeper wells.

Rotary drilling simply defines the process in which a sharp bit penetrates into the rock due to its weight and rotational movement [41]. Rotary drilling system comprises prime movers, hoisting equipment, rotary equipment, and circulating equipment, all of which mounted on a rig. The prime mover, usually a diesel engine, provides the power required for the whole rig. Hoisting system is responsible for raising and lowering the drill string in and out of the hole. Rotary equipment supports the rotation of the drill bit by transforming electrical power to rotational movement. In order to transport the cuttings to the surface and also to cool the bit, the circulation equipment provides mud flow that is directed into the drill string down to the bit and returns to surface transporting the debris accumulated in the bottom of the hole.

One of the important factors in drilling process is rate of penetration, which is usually measured in terms of meter per minute or foot per minute. This parameter shows how fast the drilling process has been done, and thus, how much cost has been reduced. Through the survey of previous studies, a series of parameters were identified as having significant effect on rate of penetration during drilling operations. These parameters include rotation speed of the bit, weight on the bit, shut-in pipe pressure, mud circulation rate, yield point and plastic viscosity of the mud, and mud gel strength. In the following, each parameter is briefly described.

Bit rotation speed: In a drilling process, the bit is rotated using rotary table or top drive system. The rotation of the bit is usually measured in rotation per minute (rpm).

Weight on the bit: In order to provide the required downward force for penetrating into the rock, several drill collars are installed before the bit. The parameter is generally called weight on bit (WOB) and measured in thousand of pounds (Klb).

Standpipe pressure: Standpipe pressure (SPP) refers to the total pressure loss due to fluid friction. In detail, SPP is the summation of pressure losses in drill string, annulus, bottom hole assembly, and across the bit. The unit for measuring the SPP is pounds per square inch (psi).

Mud flow rate: In order to lubricate and cool down the bit under drilling process, a mixture of additives mixed in water or oil, which, respectively, are called water-based and oil-based drilling mud, is pumped through the drill pipe down to the bit. Drilling mud also cleans up the bottomhole by transporting the cuttings up to the surface. It also helps penetration rate as it passes bit nozzles and penetrates the rock as a water jet system. Mud flow rate is often expressed in gallons per minute (gpm).

Mud yield point: Yield point, which is usually expressed in lbf/100 ft^2, is an indicator for determining the resistance of a fluid to movement. It is a parameter of Bingham plastic model, which is equal to shear stress at zero shear rate. As attractive force among the colloidal particle increases, the mud needs more force to move; hence the yield point is considered higher.

Mud plastic viscosity: Plastic viscosity of the mud is determined by the slope of the shear stress vs. shear rate plot. Higher plastic viscosity indicates more viscous fluid and vice versa. The unit for measurement of plastic viscosity is centipoises.

Mud gel strength: Gel strength is the term that defines the shear stress measured at low shear rate after the drilling mud has been static for a certain period of time, which is 10 s and 10 min in API standard. It indicates ability of the drilling mud to suspend drill solid and weighting material when circulation is ceased. It is measured in lbf/100 ft^2 in petroleum engineering applications.

5.3 Case study

In the present study, a data set obtained from a drilling process in a gas field located in the south of Iran was used. The depth of the well was 4235, which was drilled with one run of roller-cone bit and three runs of PDC bit. The IADC code of the roller-cone bit was 435 M, and PDC bits had codes of M332, M433, and M322. Roller-cone bit was used for about 20% and PDC bits for 80% of the drilled depth. In detail, roller-cone bit was used for the depth interval of 1016–1647 m, PDC (M332) was used for depth interval of 1647–2330 m, PDC (M433) was used for depth interval of 2330–3665 m, and finally, the depth between 3665 and 4235 m was drilled by PDC (M322).

The data set consists of 3180 samples, which were taken every 1 meter of penetration from 1016 to 4235 m. The recorded variables included well depth (D), rotation speed of bit (N), weight on bit (WOB), shut-in pipe pressure (SPP), fluid rate (Q), mud weight (MW), the ratio of yield point to plastic viscosity (Yp/PV), and the ratio of 10 min gel strength to 10 s gel strength (10MGS/10SGS). The statistical summary of the data points is gathered in **Table 1**.

Parameter (unit)	Minimum value	Maximum value	Mean value
Well depth (m)	1016	4235	2636
Rotation speed of bit (rpm)	91.38	192.00	150.72
Weight on bit (Klb)	1.02	43.26	21.59
Shut-in pipe pressure (psi)	898.98	4085.82	2502.61
Fluid rate (gpm/day)	726.92	1054.75	865.17
The ratio of yield point to plastic viscosity	0.96	2.09	1.49
The ratio of 10 min gel strength to 10 s gel strength	1.13	1.50	1.27

Table 1.
Statistical summary of input data.

5.4 General description of artificial bee Colony

This algorithm was developed by Karaboga [42] and mimics the behavior of bees when they search for nectar of flowers. In a hive of bees, there are three different types of bees: scouts, employed bees, and onlookers. The scout bees start a random search of the surrounding environment in order to find flowers that secrete nectar. After finding the flowers, they keep the location in their memory. Then, they return to the hive and share their information about their findings through a process called waggle dance. Next, the other group, called employed bees, starts finding the flowers based on the information obtained from the scouts in order to exploit the nectar of the flowers. The number of employed bees is equal to number of food sources. The third group of bees are called onlookers, which remain in the hive waiting for the return of the employed bees in order to exchange information and

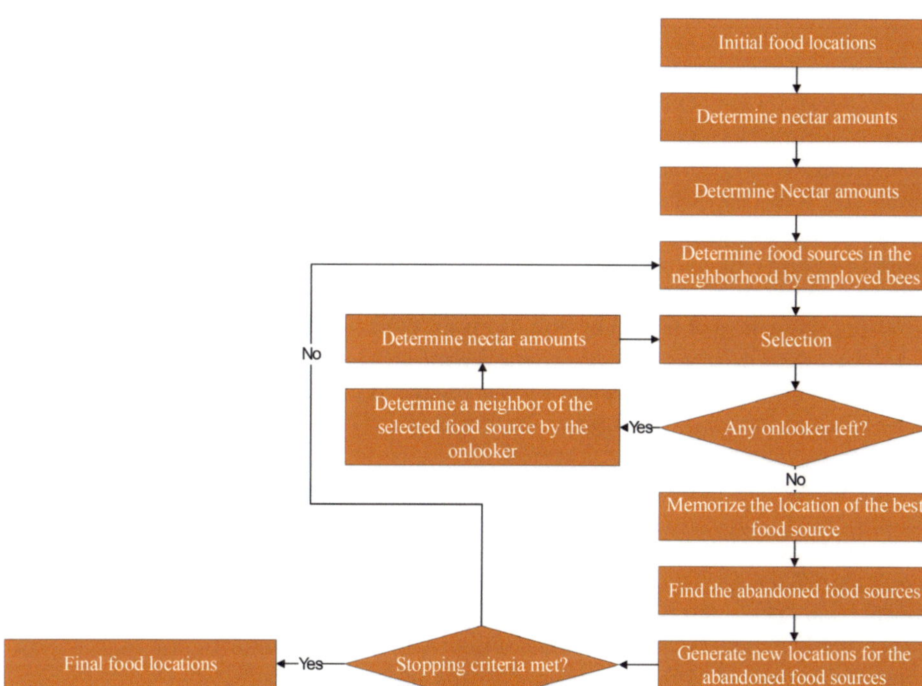

Figure 12.
Typical flowchart of ABC algorithm.

select the best source based on the dances (fitness of the candidates). In addition, the employed bees of an abandoned food site serves as a scout bee.

Considering an objective function, $f(x)$, the probability of a food source to be chosen by an onlooker can be expressed as [42]:

$$P_i = \frac{F(x_i)}{\sum_{j=1}^{S} F(x_i)} \tag{14}$$

where S indicates the number of food sources and $F(x)$ represents the amount of nectar at location x. The intake efficiency is defined as F/τ, in which τ represents the time consumed at the food source. If in a predefined number of iterations, a food source is tried with no improvement, then the employed bees dedicated to this location become scout and hence start searching the new food sources in a random manner.

ABC algorithm has been used in different engineering problems including well placement optimization of petroleum reservoirs [43], optimization of water discharge in dams [44], data classification [45], and machine scheduling [46]. More description on the ABC algorithm can be found in other references [47–50]. A typical flowchart of ABC algorithm is shown in **Figure 12**.

6. Result and discussion

6.1 Prediction

In the present research, an ANN model was developed to predict the ROP as a function of effective parameters. The neural network is widely used in various engineering fields [51–60]. In order to train the network, three training functions were used including Levenberg-Markvart (LM), scaled conjugate gradient (SCG), and one-step secant (OSS). The number of hidden layers in the network was one since according to Hornik et al. [61], one hidden layer is capable of solving any type of nonlinear function. The number of neurons in the hidden layer was another parameter to be set. Several equations have been proposed by different authors to determine the optimum number of neurons in a hidden layer, which are represented in **Table 2**. N_i and N_o indicate the number of input and output variables, respectively.

Using the values obtained by equations of **Table 2**, several ANN models were developed with neurons of 2–16. Then, the models were compared in terms of R^2 and RMSE, and the best model was selected [69, 70, 56, 71]. The comparison was done through the method proposed by Zorlu et al. [72]. In this method, the R^2 and RMSE of each enveloped model are calculated. Next, the networks are assigned an integer number according to their R^2 and RMSE value, in the way that the better result acquires higher number. For example, if the number of models is equal to 8, the model having the best (highest) R^2 value acquires 8, and the model having the worst model acquires the value of 1. This procedure also is repeated based on RMSE comparison. Then, the two numbers assigned to each model are summed up, and a total score is obtained for each model. Finally, the model acquiring the highest total value is determined as the best model for the problem of study.

In the present article, three types of learning functions were used for training the network, results of which are presented in **Tables 3–5**. According to the tables, LM, SCG, and OSS functions acquired the best results, respectively. In order to design an

Relationships	Reference
$\leq 2\times$ Ni + 1	[62]
(Ni + N0)/2	[63]
$\frac{2+N_0 \times N_i + 0.5\, N_0 \times (N_0^2 + N_i) - 3}{N_i + N_0}$	[64]
2Ni/3	[65]
$\sqrt{N_i \times N_0}$	[66]
2Ni	[67, 68]

Table 2.
The equations for determining the optimum number of neurons in a hidden layer.

Model no.	Neuron no.	Train		Test		Train rating		Test rating		Total rank
		R^2	RMSE	R^2	RMSE	R^2	RMSE	R^2	RMSE	
1	2	0.839	0.1040	0.816	0.1076	1	1	1	1	4
2	4	0.899	0.0821	0.885	0.0893	5	6	4	4	19
3	6	0.902	0.0850	0.897	0.0818	6	4	8	8	26
4	8	0.882	0.0897	0.884	0.0886	2	2	3	5	12
5	10	0.893	0.0868	0.887	0.0910	4	3	5	2	14
6	12	0.892	0.0827	0.875	0.0907	3	5	2	3	13
7	14	0.908	0.0800	0.892	0.0885	7	7	6	6	26
8	16	0.912	0.0779	0.893	0.0863	8	8	7	7	30

Table 3.
The results of the developed ANN models based on LM function.

Model no.	Neuron no.	Train		Test		Train rating		Test rating		Total rank
		R^2	RMSE	R^2	RMSE	R^2	RMSE	R^2	RMSE	
1	2	0.798	0.1159	0.824	0.1002	1	1	3	4	9
2	4	0.820	0.1092	0.815	0.1083	4	4	2	2	12
3	6	0.809	0.1127	0.839	0.0949	2	2	6	8	16
4	8	0.841	0.1035	0.831	0.0993	6	6	4	5	21
5	10	0.827	0.1076	0.846	0.0982	5	5	7	7	24
6	12	0.814	0.1093	0.810	0.1093	3	3	1	1	8
7	14	0.853	0.0984	0.837	0.1065	8	8	5	3	24
8	16	0.849	0.1006	0.860	0.0985	7	7	8	6	28

Table 4.
The results of the developed ANN models based on SCG function.

accurate model, the best model of each function was compared. The results of comparison are shown in **Figures 13** and **14**. As can be seen, the best model of LM function yielded better performance. Thus, this function was selected for designing an ANN for prediction and optimization of ROP.

Model no.	Neuron no.	Train		Test		Train rating		Test rating		Total rank
		R^2	RMSE	R^2	RMSE	R^2	RMSE	R^2	RMSE	
1	2	0.815	0.1128	0.807	0.1033	2	2	4	5	13
2	4	0.811	0.1089	0.781	0.1254	1	4	1	1	7
3	6	0.829	0.1072	0.791	0.1086	5	6	2	3	16
4	8	0.816	0.1113	0.843	0.0976	3	3	8	7	21
5	10	0.837	0.1128	0.792	0.1057	7	2	3	4	16
6	12	0.822	0.1085	0.828	0.0971	4	5	5	8	22
7	14	0.849	0.0996	0.836	0.1098	8	8	6	2	24
8	16	0.832	0.1055	0.840	0.1006	6	7	7	6	26

Table 5.
The results of the developed ANN models based on OSS function.

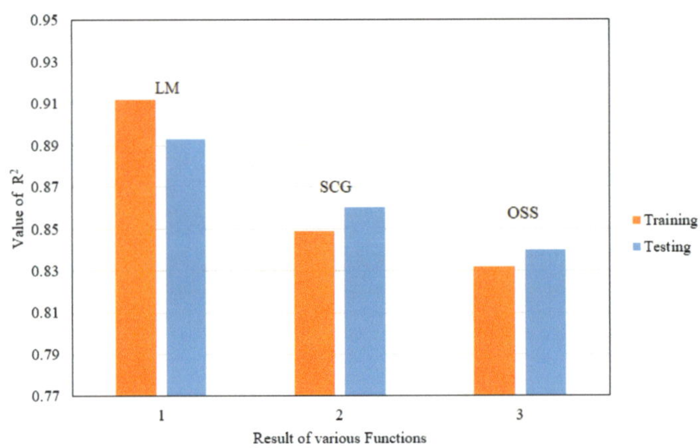

Figure 13.
The results of R^2 for LM, SCG, and OSS functions.

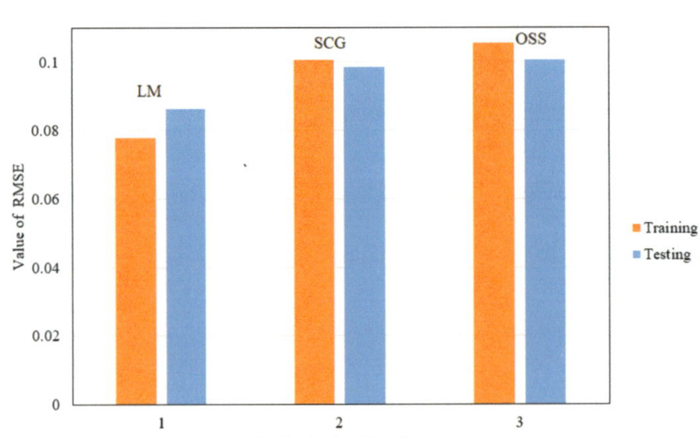

Figure 14.
The results of RMSE for LM, SCG, and OSS functions.

6.2 Optimization

In the previous section, an ANN was developed for prediction of ROP using the input data. As mentioned, selecting the most accurate predictive model can significantly affect the performance of optimization. In this section, the performance of the optimization algorithm is evaluated. Then, the ANN model obtained in the previous section is incorporated in the optimization algorithm to optimize the effective parameters for maximizing the penetration rate.

6.3 Evaluation of optimization algorithm

In this section, the best ANN model obtained in the previous section was selected for optimization of ROP using ABC algorithm. In order to evaluate the performance of ABC, two functions were used for minimization by ABC:

$$F_1(x) = \left[1 + (x_1 + x_2 + 1)^2 (19 - 14x_1 + 3x_1^2 - 14x_2 + 6x_1x_2 + 3x_2^2)\right] \times \left[30 + (2x_1 - 3x_2)^2 (18 - 32x_1 + 12x_1^2 + 48x_2 - 36x_1x_2 + 27x_2^2)\right] \quad (15)$$

The range of variations of x1 and x2 are (−2, 2). Also, the optimal value of this function at the point (1−, 0) is 3.

This function is plotted in **Figure 15**. The ABC algorithm was used for finding minimum point of the above mentioned function, and the values of −0.33559 and −0.52311 were obtained for Eq. (15). The performance of ABC in finding the minimum point is illustrated in **Figure 16**.

6.4 Optimization of ROP in petroleum wells

In this section, the ANN predictive model was used for optimization of parameters effective on ROP. Since the well depth increases during drilling, it was not considered as a decision variable. Hence, the parameters of ROP were optimized in

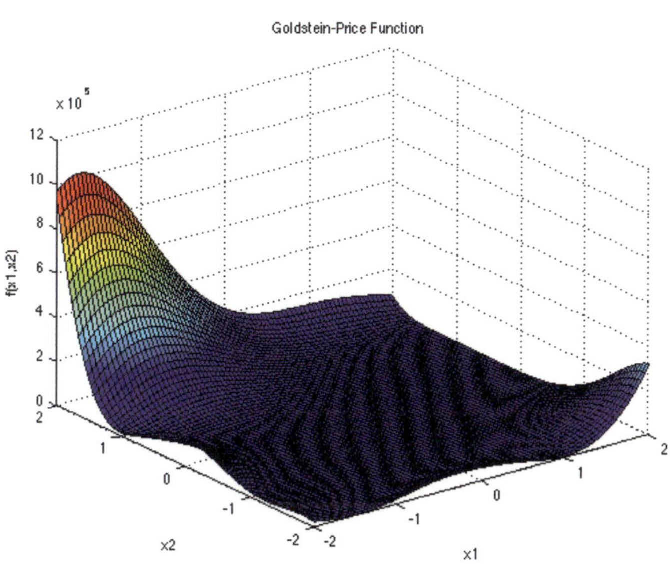

Figure 15.
Function of Eq. (6) plotted in Cartesian coordinates.

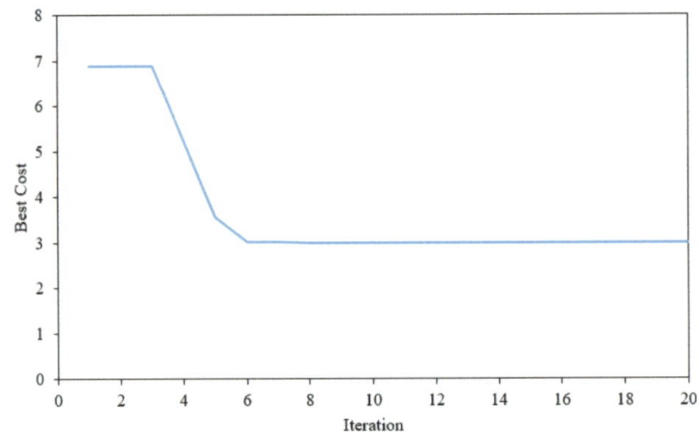

Figure 16.
Evaluation of ABC algorithm for Eq. (6).

some specific depths. It makes sense in the way that the parameters cannot be optimized in each meter of penetration.

The ABC algorithm was used for optimization of ROP effective parameters. After a series of sensitivity analysis, it was concluded that the efficient number of population and iterations are 40 and 500, respectively. Three depths on which optimization applied were 2000, 2500, and 3000. The results of optimization in the selected depths are provided in **Tables 6–8**.

As can be seen, in each selected depth, value of ROP was increased by about 20–30%. Therefore, by combining artificial intelligence and optimization, suitable

Parameter	Unit	Initial value	Optimum value
WOB	Klb	23.8	17.4
N	rpm	181	149
SPP	psi	2181.4	2783.6
Q	bbl/day	901.67	848
Y_p/PV	—	1.545	1.34
10MGS/10SGS	—	1.33	1.16
ROP	m/h	16.77	21.66

Table 6.
Comparison of real and optimized values for depth of 2000 m.

Parameter	Unit	Initial value	Optimum value
WOB	Klb	15.4	21.6
N	rpm	157	162
SPP	psi	2531.5	2481.3
Q	bbl/day	898.45	790
Y_p/PV	—	2.09	1.76
10MGS/10SGS	—	1.2	1.09
ROP	m/h	18.52	22.85

Table 7.
Comparison of real and optimized values for depth of 2500 m.

Parameter	Unit	Initial value	Optimum value
WOB	Klb	21.9	25.5
N	rpm	142	153
SPP	psi	2854.7	2927.5
Q	bbl/day	851.7	816
Y_P/PV	—	1.428	1.59
10MGS/10SGS	—	1.25	1.11
ROP	m/h	13.94	17.30

Table 8.
Comparison of real and optimized values for depth of 3000 m.

patterns for ROP in an oil well in order to increase penetration and reduce costs can be created.

6.5 Conclusion and summary

In this chapter, firstly, the basics of optimization are explained to solve problems. Then, an application of neural network combined with ABC algorithm was used for prediction of rate of penetration in a gas well. The data were collected from a gas field located in south of Iran. Seven input parameters were selected as input data to develop a predictive ANN model. For this purpose, three learning functions were compared, and LM function was selected as the best function for designing the predictive model. Next, an ABC algorithm was employed to optimize the effective parameters of ROP for maximizing the penetration rate. Three scenarios were selected for considering the well depth in optimization process. Then, the best models for the depths of 2000, 2500, and 3000 m were obtained, and the results showed 20–30% of improvement in penetration rate.

According to the results of the test, it was concluded that the proposed model is a powerful tool for prediction and optimization of rate of penetration during drilling process. Since the drilling process involves numerous effective parameters, it is almost infeasible to explicitly take into account each parameter. Therefore, use of ANN seems very useful in this complex problem and it helps to predict and optimize the penetration rate in a short period of time and without heavy computational costs.

Author details

Mohammadreza Koopialipoor[1*] and Amin Noorbakhsh[2]

1 Faculty of Civil and Environmental Engineering, Amirkabir University of Technology, Tehran, Iran

2 Department of Petroleum Engineering, Amirkabir University of Technology, Tehran, Iran

*Address all correspondence to: mr.koopialipoor@aut.ac.ir

References

[1] Azizi A. Introducing a novel hybrid artificial intelligence algorithm to optimize network of industrial applications in modern manufacturing. Complexity. 2017:18. https://doi.org/10.1155/2017/8728209

[2] Holland JH. Genetic algorithms. Scientific American. 1992;**267**:66-73

[3] Kirkpatrick S, Gelatt CD, Vecchi MP. Optimization by simulated annealing. Science (80-). 1983;**220**:671-680

[4] Glover F. Tabu search—Part I. ORSA Journal on Computing. 1989;**1**:190-206

[5] Dorigo M, Birattari M. Ant colony optimization. In: Encyclopedia of Machine Learning. Springer; 2011. pp. 36-39

[6] James K, Eberhart R. A new optimizer using particle swarm theory. In: Proceedings of the IEEE sixth international symposium on micro machine and human science.1995;**34**: 39-43

[7] Kennedy J. Particle swarm optimization. In: Encyclopedia of Machine Learning. Springer; 2011. pp. 760-766

[8] Geem ZW, Kim JH, Loganathan GV. A new heuristic optimization algorithm: Harmony search. SIMULATION. 2001; **76**:60-68

[9] Yang X-S. Metaheuristic optimization. Scholarpedia. 2011;**6**:11472

[10] Erol OK, Eksin I. A new optimization method: Big bang–big crunch. Advances in Engineering Software. 2006;**37**:106-111

[11] Yang X-S. Firefly algorithms for multimodal optimization. In: International Symposium on Stochastic Algorithms. Springer; 2009. pp. 169-178

[12] Atashpaz-Gargari E, Lucas C. Imperialist competitive algorithm: An algorithm for optimization inspired by imperialistic competition. In: Evolutionary Computation, 2007. CEC 2007. IEEE Congress on. IEEE; 2007. pp. 4661-4667

[13] Yang X-S, Deb S. Cuckoo search: Recent advances and applications. Neural Computing and Applications. 2014;**24**:169-174

[14] Yang X-S. A new metaheuristic bat-inspired algorithm. In: Nature Inspired Cooperative Strategies for Optimization (NICSO 2010). Springer; 2010. pp. 65-74

[15] Kaveh A, Talatahari S. A novel heuristic optimization method: Charged system search. Acta Mechanica. 2010; **213**:267-289

[16] Gandomi AH, Alavi AH. Krill herd: A new bio-inspired optimization algorithm. Communications in Nonlinear Science and Numerical Simulation. 2012;**17**:4831-4845

[17] Kaveh A, Farhoudi N. A new optimization method: Dolphin echolocation. Advances in Engineering Software. 2013;**59**:53-70

[18] Maurer WC. The" perfect-cleaning" theory of rotary drilling. Journal of Petroleum Technology. 1962;**14**:1-270

[19] Galle EM, Woods HB. Best constant weight and rotary speed for rotary rock bits. In: Drilling and Production Practice. Society of Petroleum Engineers: American Petroleum Institute; 1963

[20] Mechem OE, Fullerton HB Jr. Computers invade the rig floor. Oil and Gas Journal. 1965;**1**:14

[21] Bourgoyne AT Jr, Young FS Jr. A multiple regression approach to optimal

drilling and abnormal pressure detection. Society of Petroleum Engineers Journal. 1974;**14**:371-384

[22] Tansev E. A heuristic approach to drilling optimization. In: Fall Meeting of the Society of Petroleum Engineers of AIME. Society of Petroleum Engineers; 1975

[23] Al-Betairi EA, Moussa MM, Al-Otaibi S. Multiple regression approach to optimize drilling operations in the Arabian gulf area. SPE Drilling Engineering. 1988;**3**:83-88

[24] Maidla EE, Ohara S. Field verification of drilling models and computerized selection of drill bit, WOB, and drillstring rotation. SPE Drilling Engineering. 1991;**6**:189-195

[25] Hemphill T, Clark RK. The effects of pdc bit selection and mud chemistry on drilling rates in shale. SPE Drilling and Completion. 1994;**9**:176-184

[26] Fear MJ. How to improve rate of penetration in field operations. SPE Drilling and Completion. 1999;**14**:42-49

[27] Ritto TG, Soize C, Sampaio R. Robust optimization of the rate of penetration of a drill-string using a stochastic nonlinear dynamical model. Computational Mechanics. 2010;**45**: 415-427

[28] Alum MA, Egbon F. Semi-analytical models on the effect of drilling fluid properties on rate of penetration (ROP). In: Nigeria Annual International Conference and Exhibition. Society of Petroleum Engineers; 2011

[29] Yi P, Kumar A, Samuel R. Realtime rate of penetration optimization using the shuffled frog leaping algorithm. Journal of Energy Resources Technology. 2015;**137**:32902

[30] Hankins D, Salehi S, Karbalaei Saleh F. An integrated approach for drilling optimization using advanced drilling optimizer. Journal of Petroleum Engineering. 2015;**2015**:12. http://dx.doi.org/10.1155/2015/281276

[31] Asgharzadeh Shishavan R, Hubbell C, Perez H, et al. Combined rate of penetration and pressure regulation for drilling optimization by use of high-speed telemetry. SPE Drilling and Completion. 2015;**30**:17-26

[32] Wang Y, Salehi S. Application of real-time field data to optimize drilling hydraulics using neural network approach. Journal of Energy Resources Technology. 2015;**137**:62903

[33] Koopialipoor M, Armaghani DJ, Hedayat A, et al. Applying various hybrid intelligent systems to evaluate and predict slope stability under static and dynamic conditions. Soft Computing. 2018;**34**:1-17

[34] Koopialipoor M, Armaghani DJ, Haghighi M, Ghaleini EN. A neuro-genetic predictive model to approximate overbreak induced by drilling and blasting operation in tunnels. Bulletin of Engineering Geology and the Environment. 2017;**33**:1-10

[35] Hasanipanah M, Armaghani DJ, Amnieh HB, et al. A risk-based technique to analyze Flyrock results through rock engineering system. Geotechnical and Geological Engineering. 2018;**34**:1-14

[36] Koopialipoor M, Nikouei SS, Marto A, et al. Predicting tunnel boring machine performance through a new model based on the group method of data handling. Bulletin of Engineering Geology and the Environment. 2018;**34**: 1-15

[37] Bradley HB. Petroleum Engineering Handbook. Society of Petroleum Engineers; 1987

[38] Pirson SJ. Oil Reservoir Engineering. RE Krieger Publishing Company; 1977

[39] Ahmed T. Reservoir Engineering Handbook. Elsevier; 2006

[40] Bourgoyne AT, Millheim KK, Chenevert ME, Young FS. Applied Drilling Engineering. Society of Petroleum Engineers; 1986

[41] Mitchell RF, Miska SZ. Fundamentals of Drilling Engineering. Society of Petroleum Engineers; 2017

[42] Karaboga D. An Idea Based on Honey Bee Swarm for Numerical Optimization. Technical Report-tr06, Erciyes University, Engineering Faculty, Computer Engineering Department; 2005

[43] Nozohour-leilabady B, Fazelabdolabadi B. On the application of artificial bee colony (ABC) algorithm for optimization of well placements in fractured reservoirs; efficiency comparison with the particle swarm optimization (PSO) methodology. Petroleum. 2016;**2**:79-89

[44] Ahmad A, Razali SFM, Mohamed ZS, El-shafie A. The application of artificial bee colony and gravitational search algorithm in reservoir optimization. Water Resources Management. 2016;**30**:2497-2516

[45] Zhang C, Ouyang D, Ning J. An artificial bee colony approach for clustering. Expert Systems with Applications. 2010;**37**:4761-4767

[46] Rodriguez FJ, García-Martínez C, Blum C, Lozano M. An artificial bee colony algorithm for the unrelated parallel machines scheduling problem. In: International Conference on Parallel Problem Solving from Nature. Springer; 2012. pp. 143-152

[47] Koopialipoor M, Ghaleini EN, Haghighi M, et al. Overbreak prediction and optimization in tunnel using neural network and bee colony techniques. Engineering Computations. 2018;**34**:1-12

[48] Gordan B, Koopialipoor M, Clementking A, et al. Estimating and optimizing safety factors of retaining wall through neural network and bee colony techniques. Engineering Computations. 2018;**34**:1-10

[49] Koopialipoor M, Fallah A, Armaghani DJ, et al. Three hybrid intelligent models in estimating flyrock distance resulting from blasting. Engineering Computations. 2018;**34**: 1-14

[50] Ghaleini EN, Koopialipoor M, Momenzadeh M, et al. A combination of artificial bee colony and neural network for approximating the safety factor of retaining walls. Engineering Computations. 2018;**35**:1-12

[51] Azizi A, Yazdi PG, Humairi AA. Design and fabrication of intelligent material handling system in modern manufacturing with industry 4.0 approaches. International Robotics & Automation Journal. 2018;**4**:186-195

[52] Azizi A, Entessari F, Osgouie KG, Rashnoodi AR. Introducing neural networks as a computational intelligent technique. In: Applied Mechanics and Materials. Trans Tech Publications; 2014. pp. 369-374

[53] Osgouie KG, Azizi A. Optimizing fuzzy logic controller for diabetes type I by genetic algorithm. In: Computer and Automation Engineering (ICCAE), 2010 the 2nd International Conference on. Trans Tech Publications: IEEE; 2010. pp. 4-8

[54] Azizi A. Hybrid artificial intelligence optimization technique. In: Applications of Artificial Intelligence Techniques in Industry 4.0. Trans Tech Publications: Springer; 2019. pp. 27-47

[55] Azizi A, Seifipour N. Modeling of dermal wound healing-remodeling phase by neural networks. In: Computer Science and Information Technology-Spring

Conference, 2009. IACSITSC'09. International Association of. Trans Tech Publications: IEEE; 2009. pp. 447-450

[56] Koopialipoor M, Murlidhar BR, Hedayat A, et al. The use of new intelligent techniques in designing retaining walls. Engineering Computations. 2019;**35**:1-12

[57] Zhao Y, Noorbakhsh A, Koopialipoor M, et al. A new methodology for optimization and prediction of rate of penetration during drilling operations. Engineering Computations. 2019;**35**:1-9

[58] Liao X, Khandelwal M, Yang H, et al. Effects of a proper feature selection on prediction and optimization of drilling rate using intelligent techniques. Engineering Computations. 2019;**35**:1-12

[59] Azizi A, Vatankhah Barenji A, Hashmipour M. Optimizing radio frequency identification network planning through ring probabilistic logic neurons. Advances in Mechanical Engineering. 2016;**8**:1687814016663476

[60] Azizi A. RFID network planning. In: Applications of Artificial Intelligence Techniques in Industry 4.0. Springer; 2019. pp. 19-25

[61] Hornik K, Stinchcombe M, White H. Multilayer feedforward networks are universal approximators. Neural Networks. 1989;**2**:359-366

[62] Hecht-Nielsen R. Kolmogorov's mapping neural network existence theorem. In: Proceedings of the International Joint Conference in Neural Networks. IEES press; 1989. pp. 11-14

[63] Ripley BD. Statistical aspects of neural networks. In: Networks Chaos—Statistical Probabilistic Aspects. Chapman & Hall; Vol. 50. 1993. pp. 40-123

[64] Paola JD. Neural Network Classification of Multispectral Imagery. USA: Master Tezi, Univ Arizona; 1994

[65] Wang C. A Theory of Generalization in Learning Machines with Neural Network Applications. University of Pennsylvania Philadelphia, PA, USA: 1994

[66] Masters T. Practical Neural Network Recipes in C++. Morgan Kaufmann; 1993

[67] Kanellopoulos I, Wilkinson GG. Strategies and best practice for neural network image classification. International Journal of Remote Sensing. 1997;**18**:711-725

[68] Kaastra I, Boyd M. Designing a neural network for forecasting financial and economic time series. Neurocomputing. 1996;**10**:215-236

[69] Ashkzari A, Azizi A. Introducing genetic algorithm as an intelligent optimization technique. In: Applied Mechanics and Materials. Trans Tech Publications; 2014. pp. 793-797

[70] Azizi A. Applications of Artificial Intelligence Techniques in Industry 4.0. Springer; 2018

[71] Koopialipoor M, Fahimifar A, Ghaleini EN, et al. Development of a new hybrid ANN for solving a geotechnical problem related to tunnel boring machine performance. Engineering Computations. 2019;**35**:1-13

[72] Zorlu K, Gokceoglu C, Ocakoglu F, et al. Prediction of uniaxial compressive strength of sandstones using petrography-based models. Engineering Geology. 2008;**96**:141-158

Chapter 7

Design and Analysis of SMA-Based Tendon for Marine Structures

Shahin Zareie and Abolghassem Zabihollah

Abstract

A tension-leg platform (TLP), as an offshore structure, is a vertically moored floating structure, connecting to tendon groups, fixed to subsea by foundations, to eliminate its vertical movements. TLPs are subjected to various non-deterministic loadings, including winds, currents, and ground motions, keeping the tendons under ongoing cyclic tensions. The powerful loads can affect the characteristics of tendons and cause permanent deformation. As a result of exceeding the strain beyond the elastic phase of the tendons, it makes unbalancing on the floated TLPs. Shape memory alloy (SMA)-based tendons due to their superelasticity properties may potentially resolve such problem in TLP structures. In the present work, performance and functionality of SMA wire, as the main component of SMA-based tendon under cyclic loading, have been experimentally investigated. It shows a significant enhancement in recovering large deformation and reduces the amount of permanent deformation.

Keywords: tension-leg platform, shape memory alloy, tendon, the superelasticity, cyclic load

1. Introduction

Marine and offshore structures are the key elements in energy supply chains in modern communities. A wide range of offshore structures including fixed and floated platforms, particularly tension-leg platform (TLP), are used to discover, extract, and transport the fossil fuel from seas and oceans.

According to the depth of the sea or oceans, proper offshore structures are chosen. For instance, for the depth between 300 and 1500 m, TLPs are the optimized platforms. TLPs are classes of floating offshore structures, fixed by a tendon to the seabed, as presented in **Figure 1** [2]. Tendons prevent vertical movements under tough external loadings from periodic, such as day-to-day winds and currents, to nonperiodic, like hurricanes and earthquakes [3–5]. Each of those loadings can put the integrity of the TLPs at risk.

In order to prevent any instability and damage, the applied loads should be prioritized with respect to the intensity. One of the most hazardous loadings is seismic activities, which can affect the integrity directly or indirectly by generating powerful loads, called the tsunamis. In TLPs, the seismic load's effect can be transferred by tendons to the main structures; hence, any damage or residual deformation in tendons makes the whole structure unbalanced and unstable.

In order to enhance the dynamic behavior of the tendons, the ideal tendons should be able to recover the original shape after experiencing large deformation and be able to absorb the energy of seismic activities [6–10]. With respect to these desired outcomes, the shape memory alloy (SMA) is an ideal alternative to replace the conventional materials for making tendons, as shown in **Figure 2**.

An SMA is a class of smart materials with unique properties to return original shape by applying heat after removing the load, called shape memory effect (SME) or only removing loads, called superelasticity (SE). The SME and SE's state depends on the applied straining and working temperature. In both the SME and SE, the SMA is capable of dissipating the energy of the external loads. Overall, SMAs in the SE state are much popular than the SME mode, due to the simplicity of use and no need for any external source of heat. Nowadays, SMA-based applications are extensively used in many engineering applications, such as aerospace, automotive, civil infrastructure, and particularly, marine and offshore structures [6, 9–13].

The suggested SMA-based tendons are composed of SMA wires. The functionality of SMA wires under long-term cyclic loading is a crucial parameter to use for TLPs. In this study, the performance of the SMA wires is evaluated by exposing under cyclic loading.

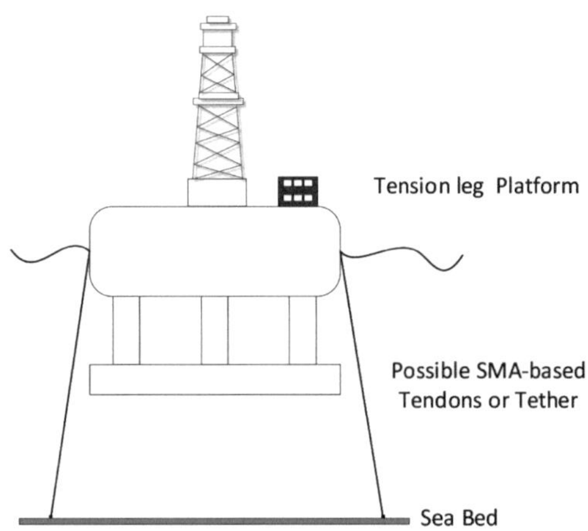

Figure 1.
The schematic diagram of TLP with SMA-based tendons taken from [1].

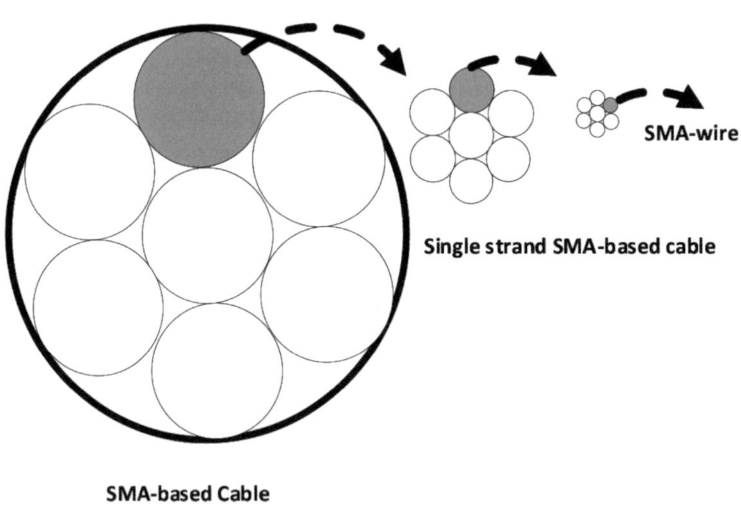

Figure 2.
The schematic diagram of the SMA-based tendon and its components taken from.

2. Shape memory alloy

The simplified constitutive law of superplastic SMA can be expressed by [14]:

$$\sigma(\varepsilon) = E\left(\varepsilon - \varepsilon^T\right) \tag{1}$$

where σ, E, ε, and ε^T are the stress, the Young modules of SMA, the strain, and the phase transformation strain, respectively.

The Young modulus of SMA is given by [14]:

$$E = E_A + \zeta(E_M - E_A) \tag{2}$$

where ξ denotes the phase transformation volume fraction. E_M and E_A are the Young modulus in the Martensite phase and the Young modulus in the Austenite phase, as presented in **Figure 3**.

In the loading phase, ε^T is given by [14]:

$$\varepsilon^T = \zeta\varepsilon_L^T \tag{3}$$

In the unloading phase, ε^T is given by

$$\varepsilon^T = \zeta\varepsilon_{ULT} \tag{4}$$

In Eqs. (3) and (4), ε_L^T and ε_{UL}^T denote the maximum phase transformation strain from Martensite to the Austenite and maximum phase transformation strain from Martensite to the Austenite from Austenite to Martensite, respectively.

The relation between ε_{UL}^T and ε_L^T is expressed by [14]:

$$\varepsilon^T{}_{UL} = \varepsilon^T{}_L + \frac{\sigma_{ms} - \sigma_{af}}{E_A} - \frac{\sigma_{mf} - \sigma_{as}}{E_M} \tag{5}$$

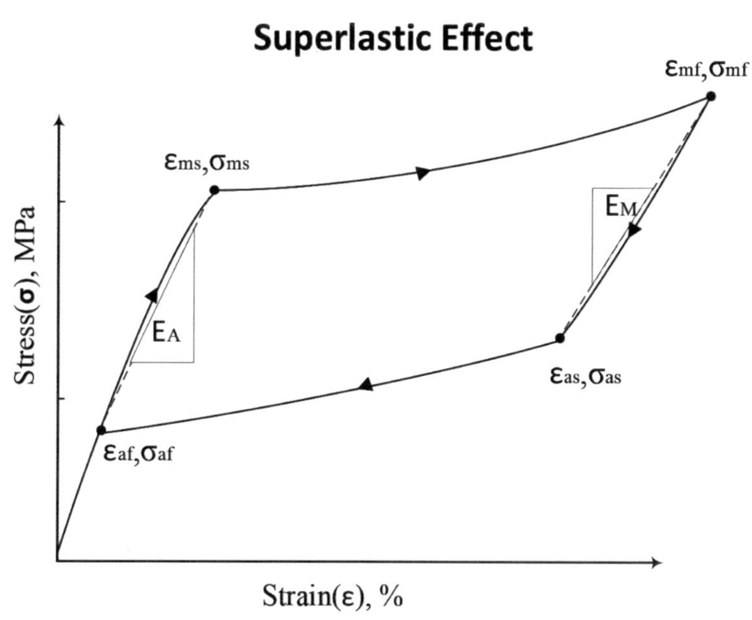

Figure 3.
The schematic diagram of strain-stress behavior of the shape memory alloy.

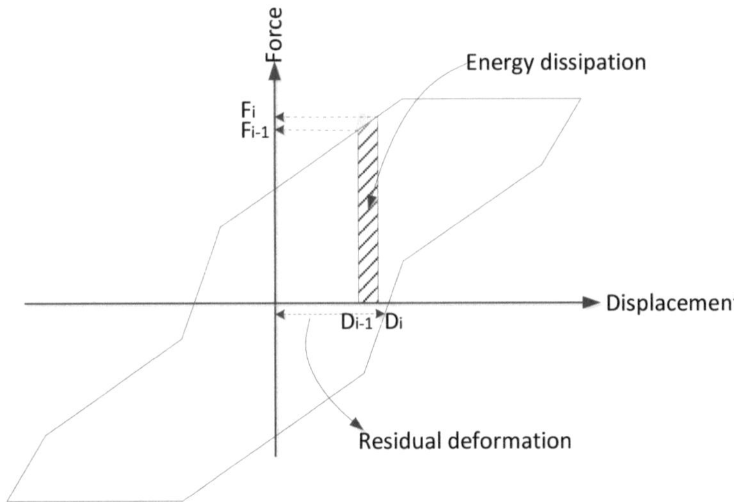

Figure 4.
The schematic diagram of the hysteresis response.

where σ_{ms} and σ_{mf} are the Martensite phase start stress to the Martensite phase finish stress, correspondingly. Similarly, σ_{as} and σ_{af} denote the Austenite phase start stress and Austenite phase finish stress. These stresses are displayed in **Figure 3**.

2.1 Energy dissipation

One of the main advantages of SMA tendon is the energy dissipation capacity. To compute this, the hysteresis response of the SMA is divided into elements, as shown in **Figure 4**. The energy dissipation of each element is expressed by

$$Energy_{total} = \sum_{i=1}^{n-1} 0.5(F_i + F_{i-1})(D_i - D_{i-1}) \tag{6}$$

where F_i and D_i stand the force and displacement of i-node in i-th element.

The total energy dissipation capacity of SMA is the sum of all energy dissipation capacity in each element.

3. Experimental configuration

In order to apply the dynamic cyclic load on the SMA wire, the MTS model 370.5 in the University of British Columbia, Okanagan Campus is used. It is a loading frame machine with an ability of 500 kN loading capacity. This machine is a programmable system equipped with sensor, actuators, control system, and software to run and collect data. This system and its accessories are displayed in **Figure 5**.

3.1 Shape memory alloy

Nowadays, among the different alloys for SMAs, nickel-titanium or Nitinol (NiTi) is one of the most common SMAs. In the present study, NiTi fabricated by Confluent Medical Technologies Company is used. The SMA specimen with

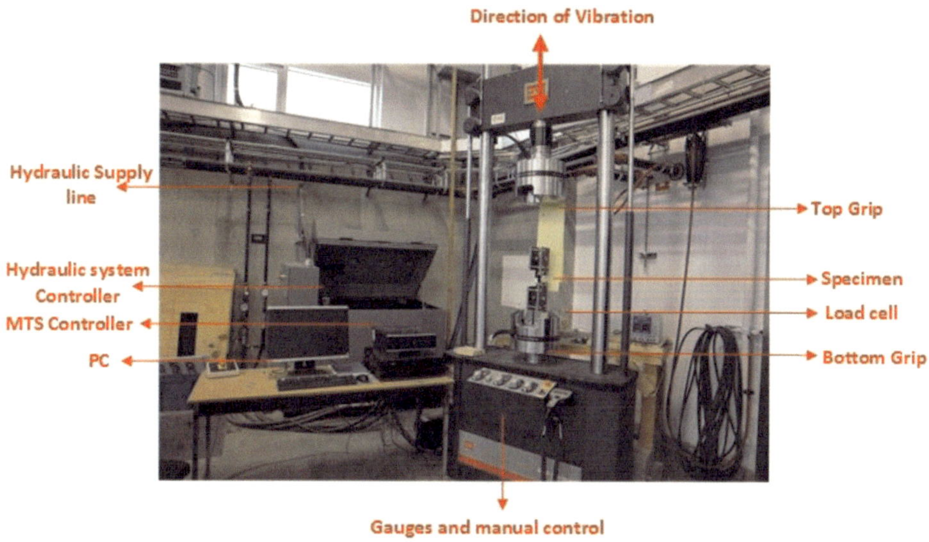

Figure 5.
The MTS model 370.5 in the UBC and its accessories.

Physical properties	Value
Melting point (°C)	1310
Density (g/cm^3)	6.5
Modulus of elasticity (GPa)	41
Coefficient of thermal expansion (\°C)	11×10^{-6}
Ultimate tensile strength (MPa)	~1070
Total elongation	~10%
Straight length (mm)	560
Radius (mm)	1.5

Table 1.
The characteristics of NiTi shape memory alloy.

0.75 mm radius and 560 mm length is kept between the top and the bottom gripper of the MTS loading frame machine by two supportive steel plates. In order to perform the experimental tests, two specimens with 0 and 1.7% applied prestrained are used.

The mechanical properties are given in **Table 1**. It is observed that the density, the melting point, the coefficient of thermal expansion, the ultimate tensile strength, and the total elongation are 1310°C, 6.5 g/cm^3, 41 GPa, ~1070 MPa, and ~10%, correspondingly.

4. Results

To simulate the effect of the long-term loading on the SMA specimens with and without applied prestraining, 1000 cyclic loads with the period of 1.4 s and the amplitude of 20 mm, as shown in **Figure 6**, are applied by the MTS loading frame machine.

The hysteresis responses of the SMA specimens are displayed in **Figure** 7. As seen, the strain-stress behavior of the SMA wires is a remarkable change under loading. It is found that the areas inside of hysteresis responses, representing energy dissipation capacities, decrease in both kinds of SMA specimens.

Another finding is the residual strain of the SMA specimen. It is noted that the residual strain of the SMA specimen without applying prestrained appears and changes up under cyclic loading. However, he prestrained SMA specimen is capable of fully recovering the original shape after exposing to the long-term cyclic loadings.

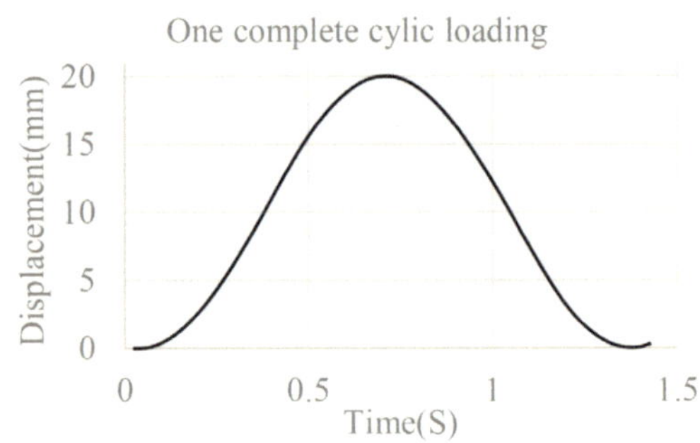

Figure 6.
1/1000 of applied cyclic loading by the MTS machine.

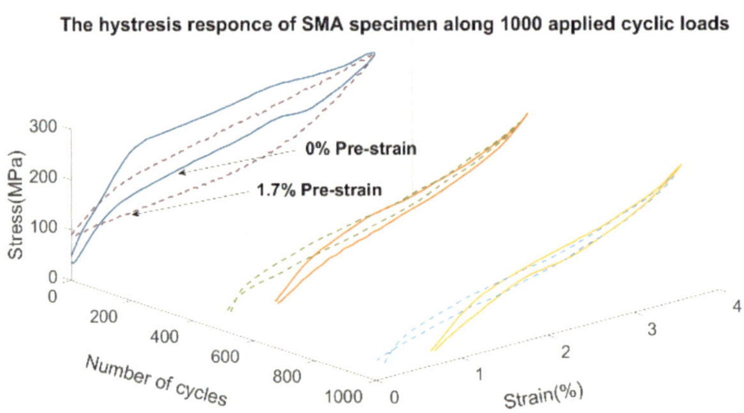

Figure 7.
Hysteresis response of the SMA along 1000 cyclic loadings.

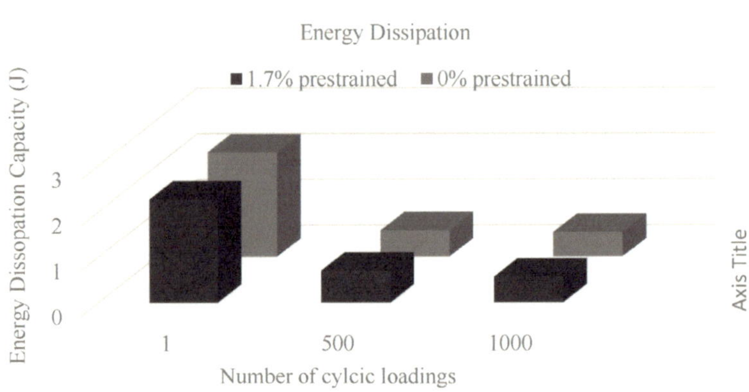

Figure 8.
The energy dissipation capacity of SMA specimens along 1000 cyclic loadings.

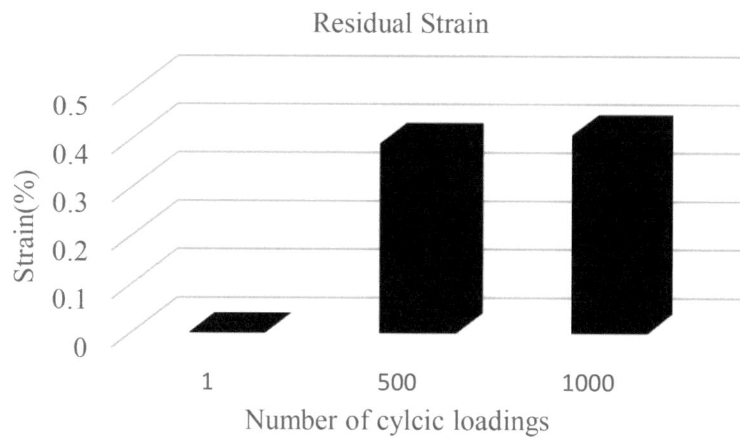

Figure 9.
The residual strain of the SMA specimen with 0% prestrained.

Figure 8 illustrates the contrast energy dissipation capacity between the SMA specimens. It is seen that the energy dissipation of the SMA without applied prestraining decreases significantly from 2.29 to 0.58 J after subjected 500 cyclic loadings and reaches 0.54 J after 1000 cyclic loadings. Under the same loading protocol, this amount in the SMA specimen with 1.7% prestrained reduces from 2.27 to 0.69 J exposed to 500 cyclic loadings and changes down to 0.56 J under 1000 cyclic loadings. This comparison shows that the rate of reduction in the energy absorption capacity of the 0% prestrained SMA wire win the first 500 cyclic is much more than the 1.7% prestrained SMA wire. Between 500 cycles and 1000 cycles, the similar drop in the energy absorption capacity in both specimens is observed.

The recovery ability is the next studied parameter in the 0% prestrained SMA specimen, as presented in **Figure 9**. Between 0-500 cyclic loading, the residual strain changes up from 0 to 2% strain of the initial strain and changes up to 3 %. As seen, a remarkable drop in the recovery ability of the SMA in the first 500 cyclic loadings. In the last 500 cyclic loadings, the steady decrease in recovery ability is observed.

5. Conclusion

This study shows that the SMA-based tendon is an ideal alternative over the conventional tendon of TLPs. It can recover the original shape up to 4% of initial length. The effect of cyclic loading on SMA wires, as the main component of the SMA-based tendons of TLP, has been examined through experimental tests.

The main outcomes of the paper are as follows:

1. The SMA wire with and without applying prestrain can absorb the remarkable energy of external excitations. However, the degradation in SMA can decrease the energy dissipation capacity under long-term loadings. Hence, at least, the safety factor of the two should be determined due to the effect of the degradation in SMAs. It covers the reduction in that capacity during long-term loadings.

2. It is also suggested to consider the residual deformation while the system is designed; another solution is to apply the restraining. This action prevents to form any residual deformation in SMA-based tendon.

For the future study, the effect of a wide range of loading's frequencies and amplitudes on the performance of the SMA-based tendons is suggested.

Acknowledgements

This research received funding from the Green Construction Research and Training Center (GCRTC) and Desmond Schumann Memorial Award. The authors wish to thank the University of British Columbia, Okanagan Campus, for the technical support. The authors also wish to thank Mr. Kyle Charles as a research engineer and ex-manager lab for his valuable help in the experimental tests at the applied laboratory for advanced materials and structures at the University of British Columbia Okanagan Campus and M. Daghighi for the help received.

Author details

Shahin Zareie[1]* and Abolghassem Zabihollah[2]

1 School of Engineering, The University of British Columbia, Kelowna, BC, Canada

2 School of Science and Engineering, Sharif University of Technology, International Campus, Kish Island, Iran

*Address all correspondence to: shahin@alumni.ubc.ca

References

[1] Gagani A, Krauklis A, Echtermeyer AT. Anisotropic fluid diffusion in carbon fiber reinforced composite rods: Experimental, analytical and numerical study. Marine Structures. 2018;**59**:47-59

[2] Bhaskara Rao DS, Panneer Selvam R. Response analysis of tension-based tension leg platform under irregular waves. China Ocean Engineering. 2016; **30**(4):603-614

[3] Guo J, Wu J, Guo J, Jiang Z. A damage identification approach for offshore jacket platforms using partial modal results and artificial neural networks. Applied Sciences. 2018;**8**(11): 2173

[4] Wang S, Li H, Han J, et al. Damage detection of an offshore jacket structure from partial modal information: Numerical study. In: The Seventh ISOPE Pacific/Asia Offshore Mechanics Symposium; 2006

[5] Asgarian B, Aghaeidoost V, Shokrgozar HR. Damage detection of jacket type offshore platforms using rate of signal energy using wavelet packet transform. Marine Structures. 2016;**45**: 1-21

[6] Zareie S, Mirzai N, Alam MS, Seethlaer RJ. A dynamic analysis of a novel shape memory alloy-based bracing system. In: CSCE 2017; 2017

[7] Zareie S, Mirzai N, Alam MS, Seethlaer RJ. An introduction and modeling of novel shape memory alloy-based bracing. In: CSCE 2017; 2017

[8] Aryan H, Ghassemieh M. A superelastic protective technique for mitigating the effects of vertical and horizontal seismic excitations on highway bridges. Journal of Intelligent Material Systems and Structures. 2017; **28**(12):1533-1552

[9] Zareie S, Alam MS, Seethaler RJ, Zabihollah A, Effect of cyclic loads on SMA-based component of cable-stayed bridge. In: 7th International Specialty Conference on Engineering Mechanics and Materials; 2019

[10] Zareie S, Alam MS, Seethaler RJ, Zabihollah A. An experimental study of SMA wire for tendons of a tension leg platform. In: 5th Annual Engineering Graduate Symposium; 2019

[11] Zareie S, Alam MS, Seethaler RJ, Zabihollah A. Effect of shape memory alloy-magnetorheological fluid-based structural control system on the marine structure using nonlinear time-history analysis. Applied Ocean Research. Oct 1 2019;**91**:101836

[12] Aryan H, Ghassemieh M. Mitigation of vertical and horizontal seismic excitations on bridges utilizing shape memory alloy system. Advanced Materials Research. 2014;**831**:90-94

[13] Mirzai N, Attarnejad R. Performance of EBFs equipped with an innovative shape memory alloy damper. Scientia Iranica. 2018. DOI: 10.24200/sci.2018.50990.1955

[14] Zuo X-B, Li A-Q, Sun X-H. Optimal design of shape memory alloy damper for cable vibration control. Journal of Vibration and Control. 2009;**15**(6): 897-921

Chapter 8

The Recent Advances in Magnetorheological Fluids-Based Applications

Shahin Zareie and Abolghassem Zabihollah

Abstract

The magnetorheological fluids (MRF) are a generation of smart fluids with the ability to alter their variable viscosity. Moreover, the state of the MRF can be switched from the semisolid to the fluid phase and vice versa upon applying or removing the magnetic field. The fast response and the controllability are the main features of the MRF-based systems, which make them suitable for applications with high sensitivity and controllability requirements. Nowadays, MRF-based systems are rapidly growing and widely being used in many industries such as civil, aerospace, and automotive. This study presents a comprehensive review to investigate the fundamentals of MRF and manufacturing and applications of MRF-based systems. According to the existing works and current and future demands for MRF-based systems, the trend for future research in this field is recommended.

Keywords: magnetorheological fluids, variable viscosity, civil, aerospace, automotive, MRF-based systems, applications

1. Introduction

The magnetorheological fluids (MRF) are a generation of smart fluids with the ability to alter their viscosity. Moreover, the state of the MRF can be switched from the semisolid to the fluid phase and vice versa upon applying or removing the magnetic field. The fast response and the controllability are the main features of the MRF-based systems, which make them suitable for applications with high sensitivity and controllability requirements. MRF-based systems are rapidly growing and widely being used in many industries such as civil, aerospace, and automotive. This chapter tends to review the fundamental concepts followed by the most recent developments in MRF-based systems. The discovery of magnetorheological fluid (MRF) goes back to 70 years ago by Rabinov [1] at the US National Bureau of Standards. Since then, hundreds of patents and research articles have been published every year.

MRF is a fluid composed of a carrier fluid, such as silicone oil and iron particles, which are dispersed in the fluid [2, 3], with an ability to alter its basic characteristics and viscosity, when subjected to the magnetic field. [4]. Upon applying a magnetic field, the tiny polarizable particles in MRF make chains between two poles, as shown in **Figure 1** [6]. The chains resist movement up to a certain breaking point (yielding point), which is a function of the strength of the magnetic field [6, 7].

IntechOpen

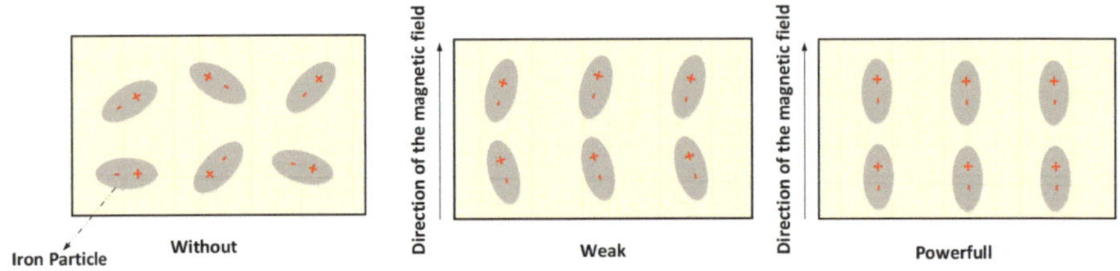

Figure 1.
The effect of magnetic field on polarization of MRFs [5].

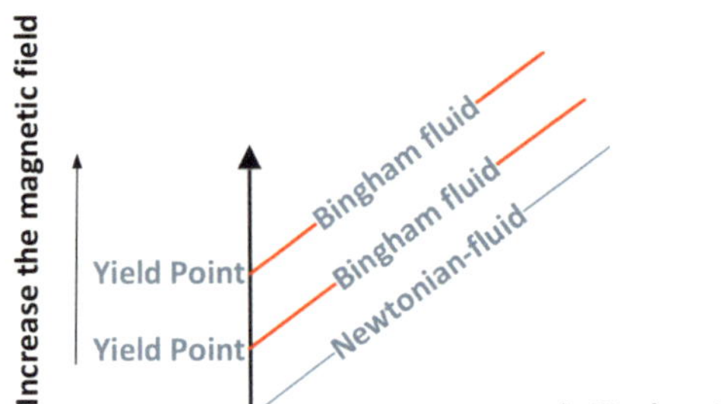

Figure 2.
The relation between shear stress and strain rate of Bingham fluid and Newtonian fluid [8].

In other words, the response of MRF is similar to non-Newtonian fluids, as shown in **Figure 2**.

2. Modeling and operational modes of MRF systems

The behavior of MRF may be described by the Bingham plasticity model [1, 2]. The model is expressed by:

$$\tau < \tau_{yield} \rightarrow \dot{y} = 0 \qquad (1)$$

$$\tau \geq \tau_{yield} \rightarrow \tau = \tau_{yield}\, \text{sgn}\,(\dot{y}) + \mu \dot{y} \qquad (2)$$

where τ_{yield}, τ, $\dot{y}$, and μ are the yielding stress, the shear stress, the strain rate, and the viscosity, respectively [1].

MRF systems operate in three basic modes, valve mode, shear mode, and squeeze mode, as shown in **Figure 3**.

In the following subsection, a brief description of each mode is provided.

2.1 MRF flow mode

The flow mode is the most common operational mode of MRF. **Figure 4** shows a simplified geometry of the flow mode. In order to obtain an in-depth understanding

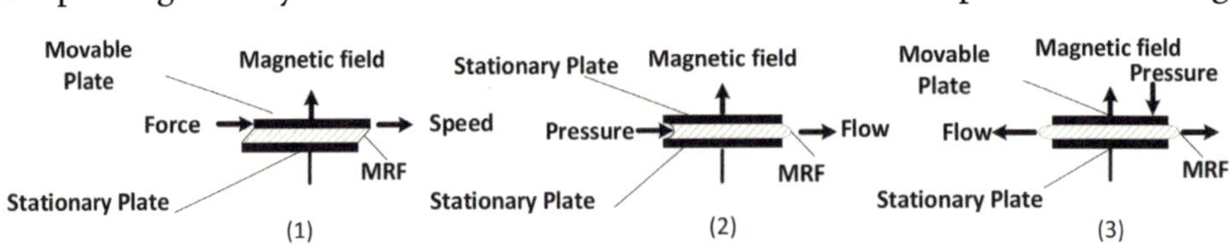

Figure 3.
The operation modes: (1) flow mode, (2) direct shear mode, (3) squeeze mode [9].

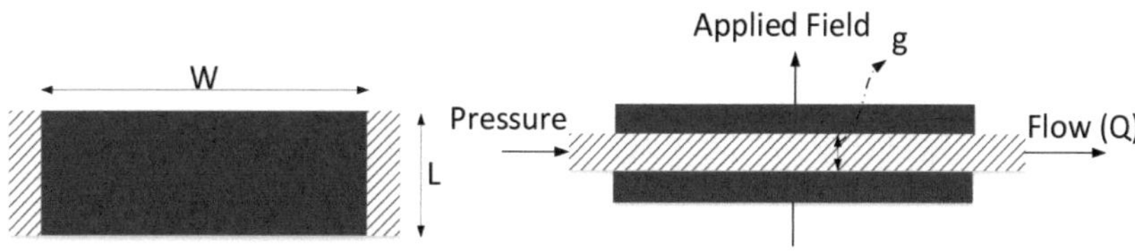

Figure 4.
The valve mode of MRF [10].

of the damping pressure supplied by MRF in this mode, one may relate the pressure due to the fluid viscosity (P_τ) and the controllable pressure (P_η).

The total damping pressure can be calculated by [10]:

$$P = P_\eta + P_\tau \tag{3}$$

where P_τ and P_τ for a Newtonian fluid are expressed by [10]:

$$P_\eta = \left(\frac{12Q\eta L}{wg^3}\right) P_\tau = \frac{C\,\tau_y L}{g} \tag{4}$$

where L and w denote the length and width of parallel plates, respectively. The term g is the gap between two plates. η and Q are the plastic viscosity and the fluid flow, correspondingly. C is a constant value and τ_y is the field-dependent yield stress.

2.1.1 MRF shear mode

The total amount of force in the shear mode between the two plates (as illustrated in **Figure 5**) is computed by [10]:

$$F = F_\eta + F_\tau \tag{5}$$

where the viscous shear force, F_η, and magnetic-dependent shear force, (F_τ), are represented by [10]:

$$F_\eta = \frac{\eta S A}{g}, \quad F_\tau = \tau_y L W \tag{6}$$

where g, A, S, and η indicate the gap between the two plates, the area of the plate, and the relative velocity between the plates, respectively. τ_y, L, and W denote the field-dependent yield stress and the width and length of the upper plate.

2.2 MRF squeeze mode

The squeeze mode occurs in two cases: compression and tension. In this study, the compressive mode of the MRF between the two plates is considered, and as a

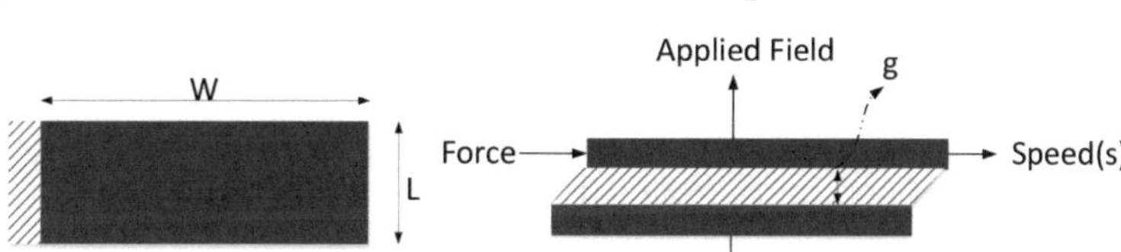

Figure 5.
The shear mode of MRF [1].

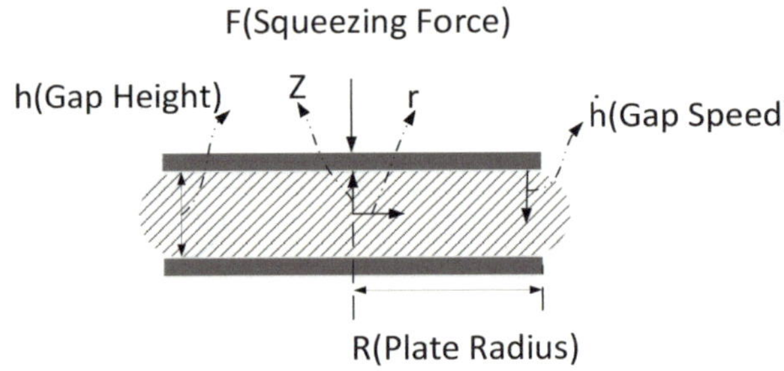

Figure 6.
The squeeze mode of the MRF [1].

result, the fluid moves between the plates as displayed in **Figure 6** [11]. The total amount of force in the squeeze mode is estimated by [11]:

$$F_s = \frac{-\pi R^4}{4}\left(\frac{6\mu\dot{h}}{h^3} + \frac{3\rho\ddot{h}}{5h} - \frac{15\rho\dot{h}^2}{14h^2}\right) \tag{7}$$

where R, h, μ, $\ddot{h}$, $\dot{h}$, and ρ are the plate radius, the distance between the two parallel plates, the viscosity of the MRF, the gap acceleration, the gap speed, and the density of the MRF, respectively.

3. MRF-based applications

Based on MRF characteristics, many devices have been developed. A summary of MRF-based devices is presented in **Figure 7**.

3.1 MRF dampers

MRF devices exhibit outstanding properties, including the large force capacity, the low voltage and the low electric current requirement, fast response, the simple interaction between the electrical current, the damping force, the adaptive rheological properties, the high viscous damping coefficient, easy controllability, and

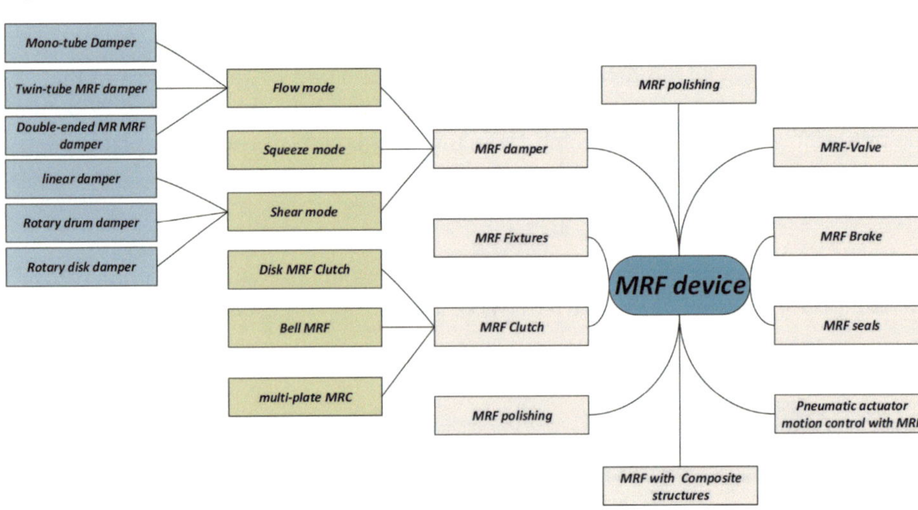

Figure 7.
The summary of MRF devices.

adaptive damping [12–16, 17–21]. MRF dampers are being used widely in the aerospace industry [22], seismic protection [2, 23], and vehicle suspension systems [24]. The core idea of designing simplified MRF dampers derived from hydraulic cylinder damper structures [10, 25]. MRF dampers have been developed based on the three basic operational modes of MRF systems.

3.1.1 MRF damper in flow mode

The most common MRF dampers are the mono-tube, the twin-tube, and the double-ended, as represented in **Figures 8–10**, respectively [26, 27]. The working principle of the mono-tube MRF damper as illustrated in **Figure 8** is based on storing pressurized gas in an accumulator located at the bottom of the damper.

Figure 9 schematically shows the working principles of a twin-tube MRF damper. The outer and inner cylinders are separated by two channels holding pressurized gas. The outer cylinder acts as an accumulator. Contrast to the mono-tube MRF damper, there are two valves: the control valve and foot valve. The function of the foot valve is to control the flow of oil to pass into the gas chamber or to extract the oil from the accumulator [10, 25, 28].

Figure 10 depicts the double-ended MRF damper, which is derived from the mono-tube MRF damper [10, 29, 30]. Two equal diameter rods are connected from the ends of the housing to the piston. It is worth noting that in double-ended MRF

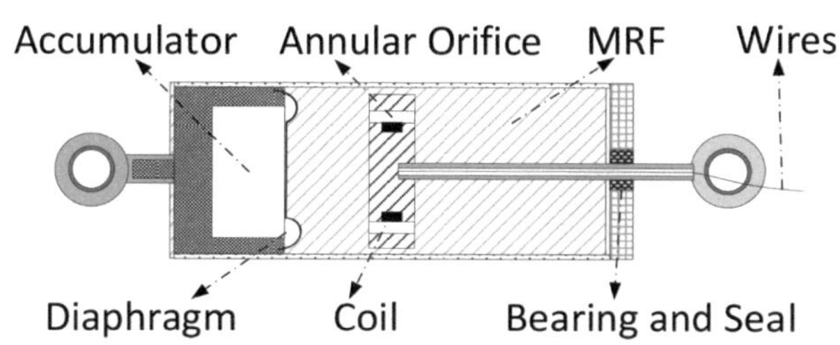

Figure 8.
Schematic diagram of mono-tube MRF damper [9].

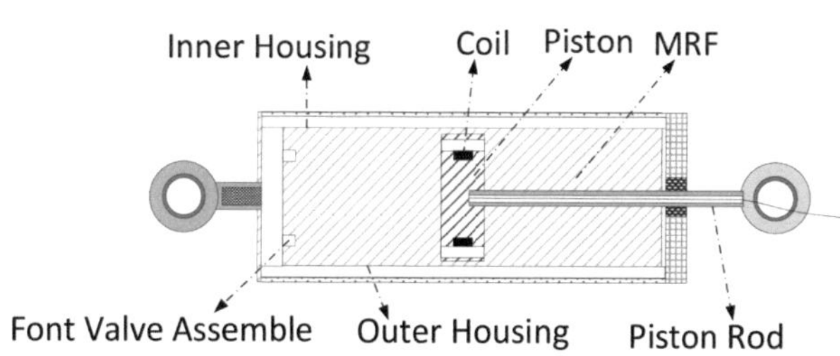

Figure 9.
Schematic diagram of the twin-tube MRF damper [9].

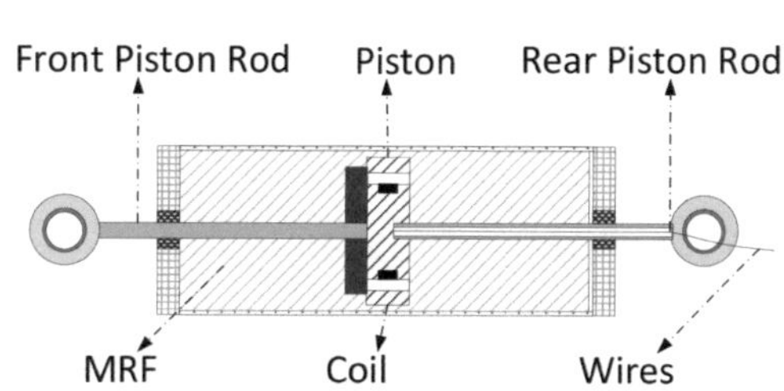

Figure 10.
Schematic diagram of the double-ended MR MRF damper [9].

damper, the accumulator is not required as long as the volume in the cylinder remains constant while the piston and rod are moving. However, sometimes a small accumulator is used for thermal expansion [30, 31].

3.1.2 MRF dampers in shear mode

Shear mode-based dampers are less common than flow mode dampers [32]. They are mostly used to damp rotational vibration. Similar to the flow mode-based MRF damper, these types of dampers work as a passive system in the absence of the magnetic field. The system can be categorized into translational linear motion, rotational disk motion, and rotational drum motion [32].

3.1.3 Linear damper

Figure 11 illustrates a linear shear mode damper which is composed of two parallel plates: a fixed plate at the bottom and a moving one at the top. The two plates are separated by a layer of MRF with thickness d [32]. The linear force (F) between the two plates can be approximated by:

$$F = \tau L b \tag{8}$$

where τ, L, and b are the shear stress, length, and width, respectively.

3.1.4 Rotary drum damper

The schematic diagram of a rotary drum damper is shown in **Figure 12**. The system is made of two concentric cylinders where the outer cylinder is held stationary and the inner cylinder rotates [32].

The damping torque (T) is computed by [32]:

$$T = 2\pi r^2 L \tau_{r\theta} \tag{9}$$

where L.r, and $\tau_{r\theta}$ are the cylinder length, the radius coordinate, and the shear stress tensor, respectively.

3.1.5 Rotary disk damper

A diagram of a rotary disk damper is shown in **Figure 13** which is composed of two disks: a fixed disk at the bottom and a rotating one at the top. The required torque to rotate the top disk is calculated by [32]:

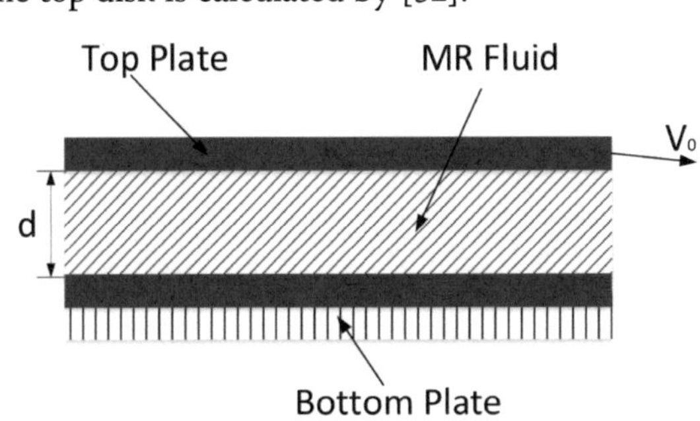

Figure 11.
The linear shear mode damper [32].

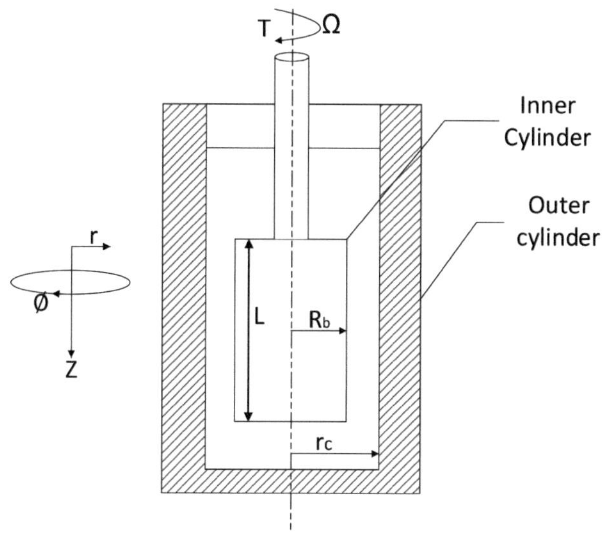

Figure 12.
The schematic diagram of rotary drum damper [32].

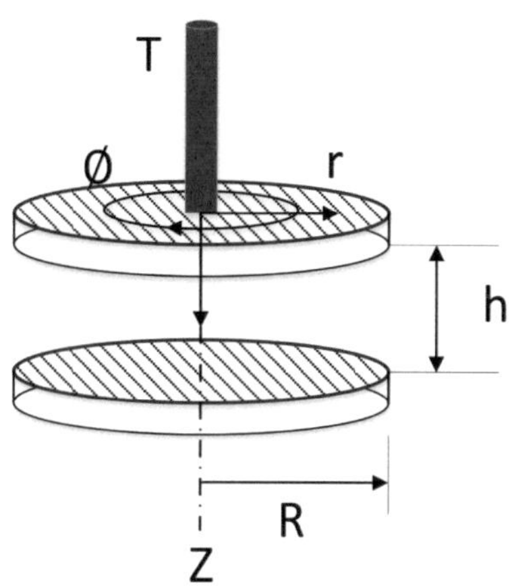

Figure 13.
The schematic diagram of rotary disk damper [32].

$$T = 2\pi \int_0^R r^2 \tau_{z\theta} dr \qquad (10)$$

where R indicates the radius of the top disk and $\tau_{z\theta}$ is the shear stress tensor in zθ plane.

The MRF damper in the shear mode provides a relative small force. However, the system prevents to form any solidification.

3.2 MRF damper in squeeze mode

Recently, due to the large force/displacement ratio, design and development of MRF dampers based on the squeeze mode have been attracted by researchers [17]. **Figure 14** shows the schematic diagram of a damper using the squeeze mode.

In contrast to the flow and the shear mode dampers, the research works in the squeeze mode dampers are very rare [17]. The total force of the top disk is calculated by [17]:

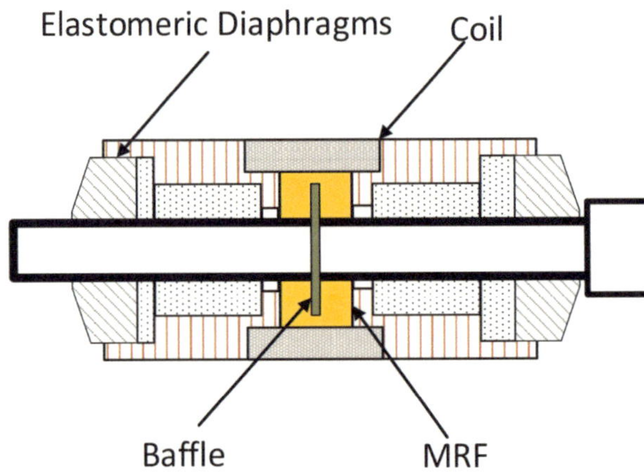

Figure 14.
Schematic diagram of MRF damper in squeeze mode [19].

$$F = 2\pi \int_{R1}^{R} rp(r)dr \qquad (11)$$

where r and $p(r)$ are the radius of the cylindrical boundary condition and the total pressure of the upper disk, respectively:

$$p(r) = p_\eta(r) + p_{MR}(r) \qquad (12)$$

where $p_\eta(r)$ and $p_{MR}(r)$ indicate the viscous pressure and the pressure considering the MRF effect, respectively. The supplied force is a function of the gap size, the MRF type, and the magnetic field intensity [17]. This damper can generate considerable damping forces while experiencing small displacements. The fundamental behavior of the MRF squeeze mode dampers is not well understood and needs to be more explored. In addition, cavitation effect needs to be considered carefully when designing squeeze mode dampers, as presented in **Table 1** [18].

3.3 MRF machining fixtures

Fixtures are important devices to precisely locate the parts during machining [20]. To respond to the demands for holding free-form parts in place efficiently, adaptive or modular fixtures have been developed. Practically, many fixtures may be required to hold all arts in desired locations [20, 21, 33].

Recently, phase-changing materials, such as MRF, become of interest in developing flexible fixtures, due to their fast response and reversibility without temperature change [20]. Due to the very low yield stress of MRF (~100 kPa), the highest clamping forces are obtained in the squeeze mode configuration [20]. **Figures 15** and **16** show two of MRF-based fixtures based on the squeeze mode developed for turbine blades.

3.4 MRF clutches

Another important MRF application is intelligent clutches [35, 36] that provide a wide torque transmissibility range upon the applied magnetic field. The long-term

System	Type	Advantages	Disadvantages
MRF damper (flow mode)	Mono-tube damper	Easy to manufacture	Required accumulator, very sensitive to any failure
	Double-ended damper	Less sensitive to failure	Required accumulator, more complex to manufacture
	Twin-tube damper	No accumulator, less sensitive	More complex to manufacture
MRF damper (squeeze mode)		Considerable damping force in small displacement	Possible cavitation, not common
MRF damper (shear mode)		Prevent solidification of MRF	Small relative force (torque)
MRF brake		Rapid response, high torque or force	High cost. Dependent upon rotational speed
MRF clutch	Disk MRC	Wide torque transmissibility range	Homogeneity of the MRF, unpredictable behavior
	Bell MRC	Wide torque transmissibility range	Homogeneity of the MRF, unpredictable behavior
	Multi-plate MRC	Faster response time, less complex, light and compact design	Self-heating, relatively high energy consumption
MRF polishing		Cause to polish surfaces smoothly	Not much effective on hard magnetic materials
MRF valve system		Less friction, fast responses, nonmoving parts, simple electrical circuit	Required extra power
Pneumatic with MRF		Stable and accurate motion control	Required extra power
MRF seals		Simple mechanism, high seal, and low maintenance required	Not effective in high rotary speed
MRF fixture		Fixing irregular-shaped objective	Not common
Composite with MRF		Adaptable damping and stiffness	Added extra weight and required extra power
MRF polishing		Cause to polish surfaces smoothly	Not much effective on hard magnetic materials

Table 1.
The summary of advantages and disadvantages of MRF-based system.

stability, short reaction time, and good controllability are the main features of MRF clutches [37]. They are promising candidates to be replaced with conventional torque converters and hydraulic starting clutches to enhance the robustness [38].

There are two types of smart clutches as illustrated in **Figure 17**: disk MRF clutches and bell MRF clutches. They are composed of a rotor, a shaft, a coil, MRF, a small gap, and input and output components [35].

In the disk shape clutch, as in all other MRF devices, there are two states: the semisolid state and the liquid state [37]. In the semisolid state, the maximum shear stress is expressed by:

$$\tau_r = \tau_{ys}\frac{r}{R_0} \tag{13}$$

where $\tau_{y,s}r$, and R are maximum shear stress, the radius of the shaft, and the radius of the disk, respectively.

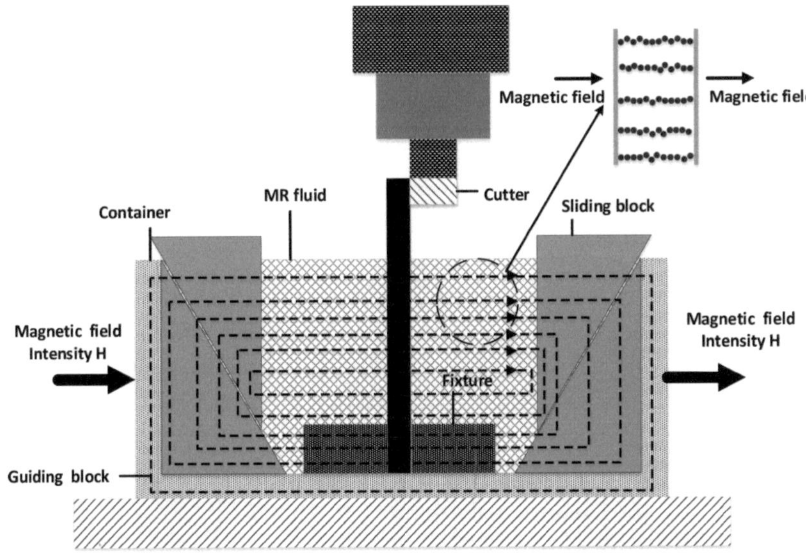

Figure 15.
The MRF-flexible-fixture prototype [34].

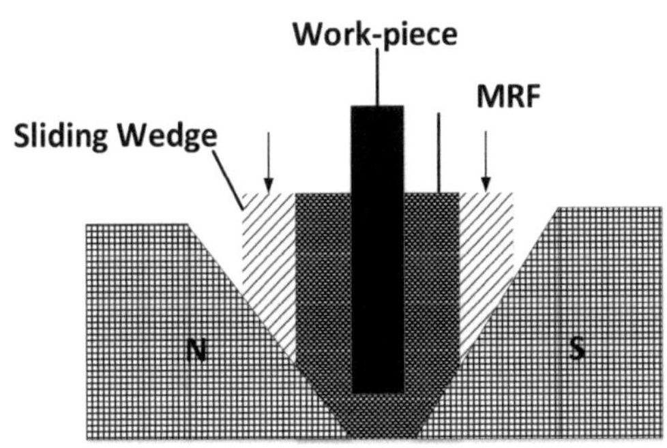

Figure 16.
The MRF-flexible-fixture prototype [20].

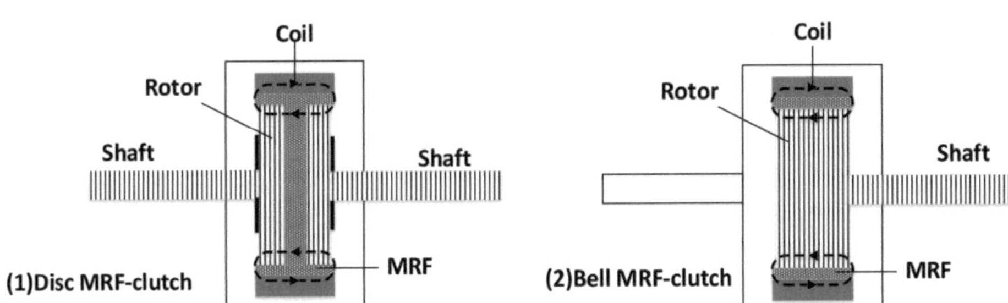

Figure 17.
Disk and bell MRF clutches [39].

$$T_{\max, S} = \pi \tau_{y,s} \frac{{R_0}^4 - R^4}{2R_0} \tag{14}$$

In the liquid state, shear stress is computed by:

$$\tau(r) = \tau_{y,d} + \eta(\omega_2 - \omega_1)\left(\frac{r}{s}\right) \tag{15}$$

where $\tau_{y,d}$, η, ω_2, and ω_1 represent the maximum shear stress, the dynamic viscosity, the angular velocity in disk 2, and angular velocity of disk 1, respectively.

The maximum torque transmitted in the liquid state in a simplified format is determined by:

$$T_{\max,L} = \pi\tau_{y,d}\left(R_0^3 - R_i^3\right)\frac{2}{3} \tag{16}$$

In the bell-shaped MRCs, the torque of the semisolid state can be given by:

$$T_{\max,S} = \left[\pi/\left(2L\tau_{(y,s)}\right)\right](R_i + R_o)^2 \tag{17}$$

where L, R_0, and R_i are the thickness of MRF, radius of input rotor, and radius of output rotor, respectively.

In the liquid state, torque is described by [37]:

$$T_{\max,L} = \left[\pi/\left(2L\tau_{(y,d)}\right)\right](R_i + R_o)^2 \tag{18}$$

The major problem in the development of this application is the difference of the density between iron particles and the carrier oil [38]. Micron-sized ferrous particles move outward faster under very high centrifugal forces. Therefore, the homogeneity of the MRF is disturbed leading to an unpredictable behavior MRC [38]. This effect has been studied by many researchers [40, 41]. In order to overcome this problem, a MRF sponge has been introduced to enhance the homogeneity of MRF in the MRC at high speeds [38].

A multi-plate MRC as shown in **Figure 18** has been introduced by Kavlicoglu et al. [42]. It is composed of 43 plates on the rotor to reduce misalignment and distribute the MRF inside the MRC more accurately [42]. Experimental and analytical works proved that the magnitude of the velocity did not affect the performance of the MRC.

The disadvantage of MRF clutches are mainly the high power required to activate the MRF and self-heating while transmitting torque from the drive side to the power off side [35].

Briefly, the comparison between different clutches is conducted and presented. It is observed that both Disk MRC and Bell MRC exhibit a wide range of torque transmissibility. It is worth noting that MRF behavior is unpredictable and the distribution of MRF is not uniform in these systems. However, multi-plate MRC is easy to manufacture, and its response is notably fast.

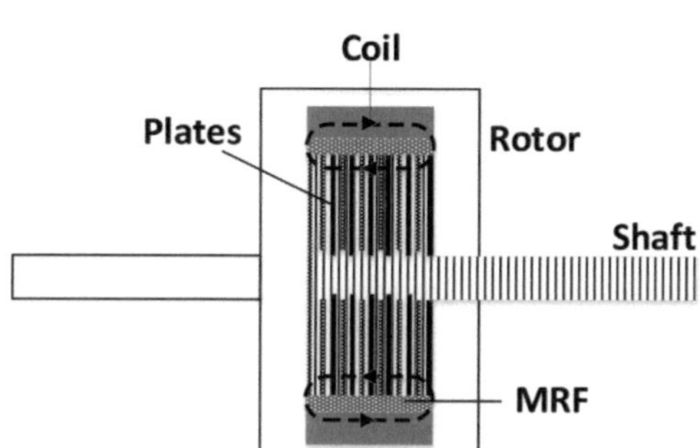

Figure 18.
Multi-plate MRC [39].

3.5 MRF polishing

Another application of MRFs is polishing or finishing based on the magnetic-assisted hydrodynamic polishing [19]. This application can be applied to plastics, optical glasses, ceramics, and complex optical devices, such as spheres. MRF polishing typically provides less surface damage compared to the conventional method [43].

Figures 19 and **20** depict the operational mechanism of the MRF polishing. As shown, the MRF fills the small gap between the workpiece and the moving wall. The magnetic field changes the viscosity and transforms it into the semisolid state [19]. The moving wall causes a profile of shear stress through the MRF layer resulting in polishing the surface of the work piece [45]. The removal rate (R) can be expressed by [43]:

$$R = KPV \tag{19}$$

where K, P, and V are the Preston coefficient, the pressure, and the velocity between the work piece and the MR fluid and the materials' removal rate, respectively [46].

The positive and negative consequences of the MRF-based polishing systems are shown in **Table 1**. It is found that the system can polish the sensitive surface smoothly. However, the system is not effective for polishing the solid magnetic surface.

3.6 MRF valves

One of the novel applications of MRF is the MRF-based valves [47–49], particularly small-size valves [50]. **Figure 21** exhibits the MRF-based valve schematically proposed by Imaduddin et al. [50].

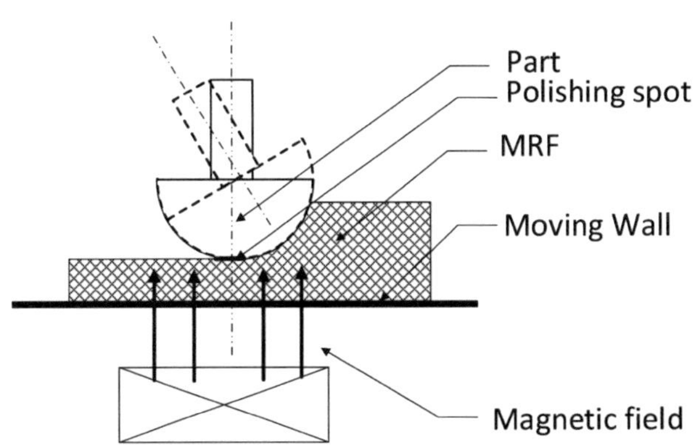

Figure 19.
The schematic diagram of MRF polishing device [44].

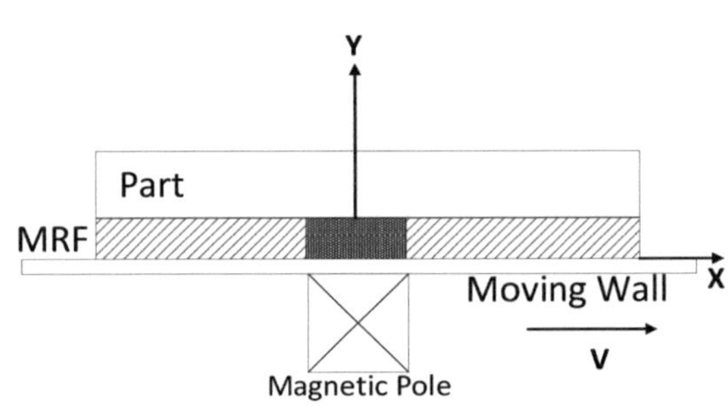

Figure 20.
Schematic working mechanism of MRF polishing [19].

Three structural configurations of MRF valves are annular, radial, and mixed annular and radial gaps [50]. **Figure 21** illustrates a mixed annular and radial gap MRF valve. The pressure drop (Δp) of the MRF valve is expressed by [50]:

$$\Delta P = \Delta P_{viscous} + \Delta P_{yield} \tag{20}$$

The pressure drop has two parts: the pressure drop $(p_{viscous})$ due to fluid viscosity and pressure drop $\left(\Delta p_{yeild}\right)$ from field-dependent yield stress [50]. $(\Delta p_{viscous})$ is computed by:

$$\Delta P_{viscous} = \frac{6\eta QL}{\pi R d^3} \tag{21}$$

where Q, L, d, and R represent the base fluid viscosity, the flow rate, the annular channel length of the valve, the valve gaps, and the channel radius, respectively.

Δp_{yeild} is computed by:

$$\Delta P_{yeild} = \frac{c\tau(\mathrm{B})\mathrm{L}}{d} \tag{22}$$

where the coefficient $(\tau(B))$ represents the field-dependent yield stress value, L is the annular channel length, and d is the gap size. The flow-velocity profile c is written by [50]:

$$c = 2.07 + \frac{12Q\eta}{12Q\eta + 0.8\pi R d^2 \tau(B)} \tag{23}$$

The strengths and weaknesses of the MRF valve system are illustrated in **Table 1**. It is observed that the MRF valves provide fast response, less friction, and simple electrical circuit for actuation.

3.7 The MRF brake

MRFs are also used to develop the new type of the braking system and can be replaced with conventional systems. The MRF brake has a high potential to decrease the transmitted torque rapidly subjected to external magnetic fields [51]. In an MRF brake system, the MRF is located between the outer cylinder and the inner rotating cylinder [19]. By energizing solenoid coil, the MRF supplies the resistance shear

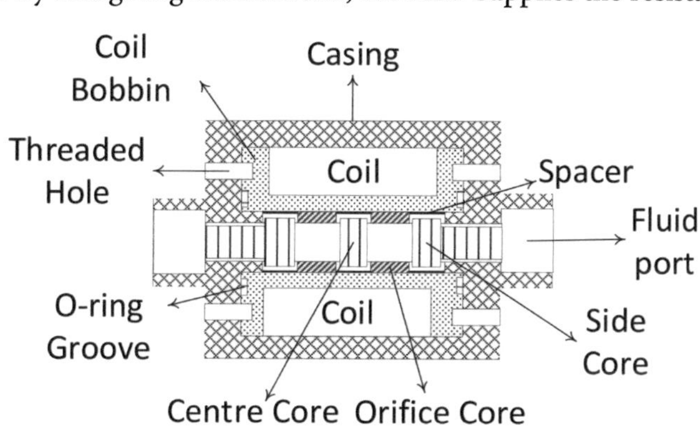

Figure 21.
Schematic diagram of the MRF valve with annular and radial gaps [50].

force in milliseconds against the torque of the shaft. By removing the magnetic field, the inner cylinder rotates freely [52]. The schematic diagram of the MRF brake is presented in **Figure 22**. The MRF brakes are available in various different shapes, such as drums, disks, and T-shaped rotors [19]. Recently, Sukhwani et al. [53] proposed a new type of the brake based on MR grease. However, their proposed brake provided lower breaking capacity than that of the existing MRF breaks.

The MRF brake system has the capability to supply a huge amount of force (torque).

3.8 MRF seals

The sealing of machinery, such as vacuuming equipment, is a significant challenge in the industry [19, 44]. The MRF is considered as a potential technology for sealing pressures up to 3300 kPa [54, 55]. Kanno et al. [54] suggested a one-step seal for a rotary shaft, as illustrated in **Figure 23** schematically. The system was tested at a rotational speed of 1000 rpm with two different sizes of gaps (1–1.7, 0.06–0.5 mm). The major benefits of the system are its ease of operation, good sealing capacity, and low maintenance requirements. Kordonsky et al. [56] studied different intensities of the magnetic fields for different shaft rotation speeds. The study showed that critical pressure is proportionally related to the square of the applied magnetic field strength. Fujita et al. [57] showed that the burst pressure of the seal is a function of size and the volume fraction of MRFs.

As displayed in **Table 1**, the MRF seal needs the external electric power to be actuated, and the performance is not efficient in the rotary movements. However, the working mechanism of MRF seal is simple with a low maintenance.

3.9 Pneumatic motion control with the MRF technology

One of the major challenges in pneumatics systems is the accuracy of servo motion control due to the high compression of air, as the working fluid [58]. There are two conventional methods including the airflow regulation and the pneumatic braking for motion control. However, the complexity of these systems is a major challenge. Recently, MRF brakes are being used to enhance the motion control of pneumatic actuators (PAMC), as displayed in **Figure 24** [44]. The system is composed of a pneumatic actuator in parallel with an MRF brake to improve the system performance and functionality due to directional control and complexity of servo

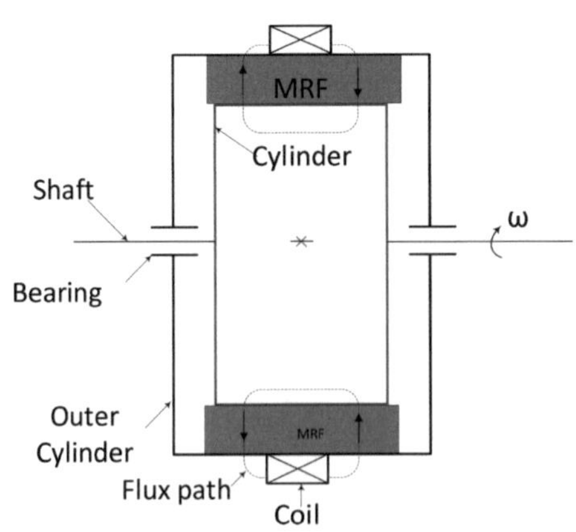

Figure 22.
The typical MRF brake [51].

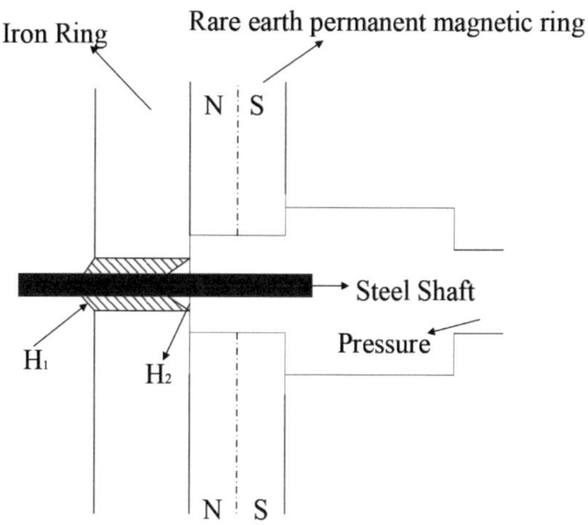

Figure 23.
Schematic diagram of one-step MRF seal [19].

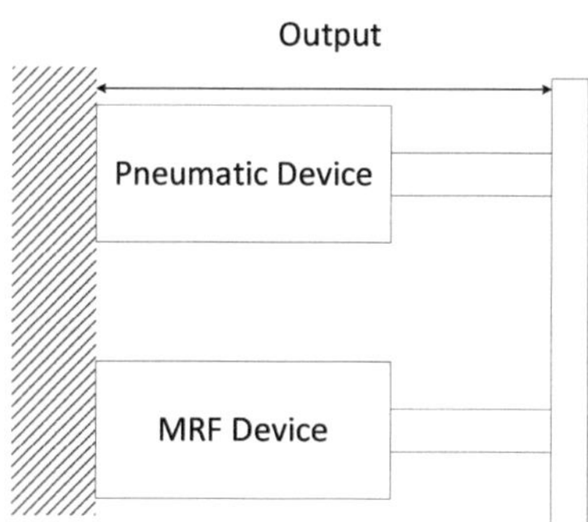

Figure 24.
Schematic diagram of MRF pneumatic motion control [58].

mechanism [58]. Moreover, the MRF can be used as the pneumatic rotary actuator to control rotary motion and velocities [44, 59–61, 74].

The pneumatic with MRF-based control movement provides higher accuracy. However, the system needs the external power for activation.

3.10 MRFs embedded in composite structures

Composite structures are gaining interest in many industries, including civil, transportation, and aerospace due to their excellent mechanical properties, particularly the strength to weight ratio [62–68]. In many applications, the composite structures are exposed to excessive vibration resulting to instability and unpredicted failure. In order to suppress the vibration in composite structures, different methods including passive, semi-active, and active vibration controls have been developed [41, 69–72].

Naji et al. [73] studied the dynamic behavior of a laminate composite beam integrated with an MRF layer, as shown in **Figure 25**. The study showed that magnetic fields in the range of 0–1600 Gauss reduce the maximum displacement and increase the natural frequency. The MRF composite has potential applications in aerospace, civil infrastructures, and automobile industries to suppress the excessive vibrations and\or control the sound propagation, as presented in **Table 1** [69].

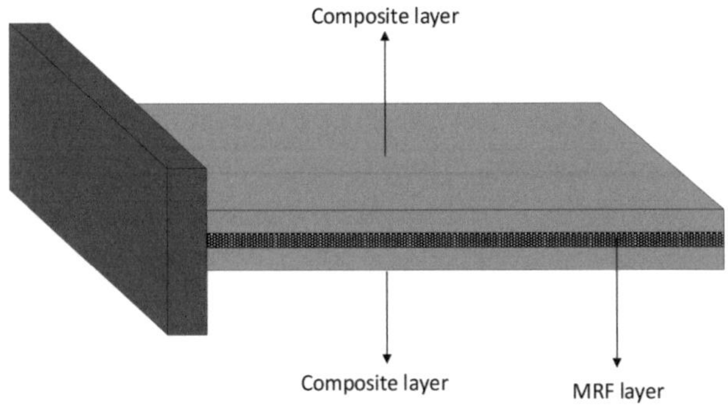

Figure 25.
Schematic diagram of an adaptive MRF-laminated composite beam [73].

It is noted that adding MRF to laminated composite structures slightly increases the weight of the element.

4. Chapter summary

In this paper, the basic knowledge of MRF and its spectacular characteristics particularly the switching phases between the semisolid state and fluid state via changing the viscosity of MRF has been concisely discussed.

According to the existing works, MRF has been found to be an excellent candidate to be replaced with the conventional fluid in the fluid-based systems. In brief, MRF-based systems improve the performance and functionality of control systems for many applications, particularly in the followings aspects:

1. Controllability: MRF-based systems provide precise output control due to the variable viscosity of MRF and switching between the semisolid and the fluid phases upon application of the magnetic field.

2. Fast response: reaction of MRF-based systems to the applied magnetic field is in the scale of milliseconds, thus making them suitable candidates to be used for real-time control applications.

3. Extensive applications: MRF-based control systems have found extensive applications in a wide range of industries, including civil, aerospace, and automobiles to enhance the performance and functionality of the systems to achieve the desired outputs.

Acknowledgements

Help received from M. Daghighi is much appreciated.

Author details

Shahin Zareie* and Abolghassem Zabihollah
School of Science and Engineering, Sharif University of Technology, International Campus, Iran

*Address all correspondence to: zareie_shahin@alum.sharif.edu

References

[1] Rabinow J. Themagnetic fluid clutch. AIEETrans. 1948;**67**:1308-1315

[2] Iqbal MF. Application of magneto-rheological dampers to control dynamic response of buildings. Concordia University; 2009

[3] Chu SY, Soong TT, Reinhorn AM. Active, Hybrid and Semi-Active Structural Control—A Design and Implementation Handbook. John Wiley & Sons; 2005. ISBN: 978-0-470-01352-6

[4] Yazid IIM, Mazlan SA, Kikuchi T, Zamzuri H, Imaduddin F. Design of magnetorheological damper with a combination of shear and squeeze modes. Materials and Design. 2014;**54**: 87-95

[5] Sapiński B, Filuś J. Analysis of parametric models of MR linear damper. Journal of Theoretical and Applied Mechanics. 2003;**41**(2):215-240

[6] Esteki K. Developing New Analytical and numerical models for Mr Fluid dampers and their application to seismic design of buildings. The Department of Building, Civil, and Environmental Engineering, Concordia University; 2014

[7] Yang S, Li S, Wang X, Gordaninejad F, Hitchcock G. A hysteresis model for magneto-rheological damper. International Journal of Nonlinear Sciences and Numerical Simulation. 2005;**6**(2):139-144

[8] Grunwald A. Design and Optimization of Magnetostrictive Actuator. Dublin City University; 2007

[9] Xiaocong Z, Xingjian J, Li C. Magnetorheological fluid dampers: A review on structure design and analysis. Journal of Intelligent Material Systems and Structures. 2012;**23**(8):839-873

[10] Poynor JC. Innovative Designs for Magneto-Rheological Dampers. Virginia Polytechnic Institute and State University. Virginia Tech; 2001

[11] McIntyre EC. Compression of Smart Materials: Squeeze Flow of Electrorheological and Magnetorheological Fluids. ProQuest; 2008

[12] Sternberg A, Zemp R, de la Llera JC. Multiphysics behavior of a magneto-rheological damper and experimental validation. Engineering Structures. 2014;**69**:194-205

[13] Luu M, Martinez-Rodrigo MD, Zabel V, Könke C. Semi-active magnetorheological dampers for reducing response of high-speed railway bridges. Control Engineering Practice. 2014;**32**:147-160

[14] Nguyen Q-H, Choi S-B. Optimal design of a vehicle magnetorheological damper considering the damping force and dynamic range. Smart Materials and Structures. 2008;**18**(1):15013

[15] De Vicente J, Klingenberg DJ, Hidalgo-Alvarez R. Magnetorheological fluids: A review. Soft Matter. 2011;7(8): 3701-3710

[16] Jiang Z, Christenson R. A comparison of 200 kN magneto-rheological damper models for use in real-time hybrid simulation pretesting. Smart Materials and Structures. 2011; **20**(6):65011

[17] Farjoud A, Cavey R, Ahmadian M, Craft M. Magneto-rheological fluid behavior in squeeze mode. Smart Materials and Structures. 2009;**18**(9):95001

[18] Boelter R, Janocha H. Design rules for MR fluid actuators in different

working modes. In: Smart Structures and Materials. Passive Damping and Isolation. International Society for Optics and Photonics; 9 May, 1997;**3045**: 148-160

[19] Hajalilou A, Mazlan SA, Lavvafi H, Shameli K. Magnetorheological Fluid Applications. In: Field Responsive Fluids as Smart Materials. Springer; 2016. pp. 67-81

[20] Tang X, Zhang X, Tao R. Flexible fixture device with magneto-rheological fluids. Journal of Intelligent Material Systems and Structures. 1999;**10**(9): 690-694

[21] Trappey JC, Liu CR. A literature survey of fixture design automation. International Journal of Advanced Manufacturing Technology. 1990;**5**(3): 240-255

[22] Batterbee DC, Sims ND, Stanway R, Wolejsza Z. Magnetorheological landing gear: 1. A design methodology. Smart Materials and Structures. 2007;**16**(6): 2429

[23] Dominguez A, Sedaghati R, Stiharu I. Modeling and application of MR dampers in semi-adaptive structures. Computers and Structures. 2008;**86**(3): 407-415

[24] Carlson JD. MR fluids and devices in the real world. International Journal of Modern Physics B. 2005;**19**(07n09): 1463-1470

[25] Gillespie T. Development of semi-active damper for heavy off-road military vehicles. Masters Abstracts International. 2006; **45**(04)

[26] Reichert BA. Application of Magnetorheological Dampers for Vehicle Seat Suspensions. Virginia Polytechnic Institute and State University. Virginia Tech; 1997

[27] Ebrahimi B. Development of Hybrid Electromagnetic Dampers for Vehicle Suspension Systems. University of Waterloo; 2009

[28] Goncalves FD. Characterizing the behavior of magnetorheological fluids at high velocities and high shear rates. Virginia Polytechnic Institute and State University; 2005

[29] Norris JA, Ahmadian M. Behavior of magneto-rheological fluids subject to impact and shock loading. In: ASME 2003 International Mechanical Engineering Congress and Exposition; 2003. pp. 199-204

[30] Abu-Ein SQ, Fayyad SM, Momani W, Al-Alawin A, Momani M. Experimental investigation of using MR fluids in automobiles suspension systems. Research Journal of Applied Sciences, Engineering and Technology. 2010;**2**(2):159-163

[31] Wang Q, Ahmadian M, Chen Z. A novel double-piston magnetorheological damper for space truss structures vibration suppression. Shock and Vibration. 2014;**2014**

[32] Wereley NM, Cho JU, Choi YT, Choi SB. Magnetorheological dampers in shear mode. Smart Materials and Structures. 2007;**17**(1):15022

[33] Arzanpour S, Fung J, Mills JK, Cleghorn WL. Flexible fixture design with applications to assembly of sheet metal automotive body parts. Assembly Automation. 2006;**26**(2):143-153

[34] Ma J, Zhang D, Wu B, Luo M, Liu Y. Stability improvement and vibration suppression of the thin-walled workpiece in milling process via magnetorheological fluid flexible fixture. The International Journal of Advanced Manufacturing Technology. 1 Feb, 2017;**88**(5-8): 1231-1242

[35] Jackel M, Kloepfer J, Matthias M, Seipel B. The novel MRF-ball-clutch design a MRF-safety-clutch for high torque applications. Journal of Physics: Conference Series. 2013;**412**(1):12051

[36] Najmaei N, Asadian A, Kermani MR, Patel RV. Magneto-rheological actuators for haptic devices: Design, modeling, control, and validation of a prototype clutch. In: Robotics and Automation (ICRA), 2015 IEEE International Conference; 2015. pp. 207-212

[37] Lampe D, Thess A, Dotzauer C. MRF-clutch-design considerations and performance. Transition. 1998;**3**:10

[38] Neelakantan VA, Washington GN. Modeling and reduction of centrifuging in magnetorheological (MR) transmission clutches for automotive applications. Journal of Intelligent Material Systems and Structures. 2005; **16**(9):703-711

[39] Xu X, Zeng C. Notice of retraction design of a magneto-rheological fluid clutch based on electromagnetic finite element analysis. In: 2nd International Conference on Computer Engineering and Technology. IEEE; 2010. Vol. 5, pp. V5-182

[40] Bansbach EA. Torque Transfer Apparatus Using Magnetorheological Fluids. Google Patents, 1998

[41] Gopalswamy S, Jones GL. Magnetorheological Transmission Clutch. Google Patents; 1998

[42] Kavlicoglu BM, Gordaninejad F, Evrensel CA, Cobanoglu N, Liu Y, Fuchs A, et al. High-torque magnetorheological fluid clutch. In: Smart Structures and Materials. International Society for Optics and Photonics. Damping and Isolation. 27 Jun 2002;**4697**:393-401

[43] Kordonski W, Golini D. Progress update in magnetorheological finishing. International Journal of Modern Physics B. 1999;**13**(14n16):2205-2212

[44] Wang J, Meng G. Magnetorheological fluid devices: principles, characteristics and applications in mechanical engineering. Proceedings of the Institution of Mechanical Engineers, Part L: Journal of Materials: Design and Applications. 2001;**215**(3):165-174

[45] Jacobs SD. Manipulating mechanics and chemistry in precision optics finishing. Science and Technology of Advanced Materials. 2007;**8**(3):153-157

[46] Jain VK, Sidpara A, Sankar MR, Das M. Nano-finishing techniques: A review. Proceedings of the Institution of Mechanical Engineers, Part C: Journal of Mechanical Engineering Science. 2012; **226**(2):327-346

[47] Manoharan V. Magneto-Rheological Fluid Device as Artificial Feel Force System On Aircraft Control Stick. Dissertations and Theses. 2016. Available from: https://commons.erau.edu/edt/225

[48] Muzakkir SM, Hirani H. A magnetorheological fluid based design of variable valve timing system for internal combustion engine using axiomatic design. International Journal of Current Innovation Research. 2015; **5**(2):603-612

[49] Yoo J-H, Wereley NM. Design of a high-efficiency magnetorheological valve. Journal of Intelligent Material Systems and Structures. 2002;**13**(10): 679-685

[50] Imaduddin F, Mazlan SA, Zamzuri H, Yazid II. Design and performance analysis of a compact magnetorheological valve with multiple annular and radial gaps. Journal of

Intelligent Material Systems and Structures. Jun 2015;**26**(9):1038-1049

[51] Huang J, Zhang JQ, Yang Y, Wei YQ. Analysis and design of a cylindrical magneto-rheological fluid brake. Journal of Materials Processing Technology. 2002;**129**(1):559-562

[52] Huang J, Wang H, Ling J, Wei YQ, Zhang JQ. Research on chain-model of the transmission mechanical property of the magnetorheological fluids. Mach. Des. Manuf. Eng. 2001;**30**(2):3-7

[53] Sukhwani VK, Hirani H. A comparative study of magnetorheological-fluid-brake and magnetorheological-grease-brake. Tribology. Online. 2008;**3**(1):31-35

[54] Kanno T, Kouda Y, Takeishi Y, Minagawa T, Yamamoto Y. Preparation of magnetic fluid having active-gas resistance and ultra-low vapor pressure for magnetic fluid vacuum seals. Tribology International. 1997;**30**(9):701-705

[55] Browne AL, Johnson NL, Barvosa-Carter W, McKnight GP, Keefe AC, Henry CP. Active Material Based Seal Assemblies. Google Patents; 2012

[56] Kordonsky W. Elements and devices based on magnetorheological effect. Journal of Intelligent Material Systems and Structures. 1993;**4**(1):65-69

[57] Fujita T, Yoshimura K, Seki Y, Dodbiba G, Miyazaki T, Numakura S. Characterization of magnetorheological suspension for seal. Journal of Intelligent Material Systems and Structures. 1999;**10**(10):770-774

[58] Jolly MR. Pneumatic motion control using magnetorheological technology. In: Smart Structures and Materials. Industrial and Commercial Applications of Smart Structures Technologies. International Society for Optics and Photonics. 14 Jun 2001;**4332**:300-308

[59] Alam M, Choudhury IA, Bin Mamat A. Mechanism and design analysis of articulated ankle foot orthoses for drop-foot. The Scientific World Journal. 2014; **2014**:14. Article ID 867869. Available from: https://doi.org/10.1155/2014/867869

[60] Zhu X, Jing X, Cheng L. A magnetorheological fluid embedded pneumatic vibration isolator allowing independently adjustable stiffness and damping. Smart Materials and Structures. 2011;**20**(8):85025

[61] Mikułowski GM, Holnicki-Szulc J. Adaptive landing gear concept feedback control validation. Smart Materials and Structures. 2007;**16**(6):2146

[62] Zareie S, Zabihollah A. A failure control method for smart composite morphing airfoil by piezoelectric actuator. Transactions of the Canadian Society for Mechanical Engineering. 2011;**35**(3):369-381

[63] Hosseiny SA, Jakobsen J. Local fatigue behavior in tapered areas of large offshore wind turbine blades. In: IOP Conference Series: Materials Science and Engineering. IOP Publishing; Jul 2016;**139**(1):012022

[64] Pol MH, Zabihollah A, Zareie S, Liaghat G. Effects of nano-particles concentration on dynamic response of laminated nanocomposite beam/Nano daleliu koncentracijos itaka dinaminei nanokompozicinio laminuoto strypo reakcijai. Mechanika. 2013;**19**(1):53-58

[65] Fattahi SJ, Zabihollah A, Zareie S. Vibration monitoring of wind turbine blade using fiber bragg grating. Wind Engineering. 2010;**34**(6)

[66] Zabihollah A, Momeni S, Ghafari AS. Effects of nanoparticles on the improvement of the dynamic response of nonuniform-thickness laminated

composite beams. Journal of Mechanical Science and Technology. 1 Jun 2016;**30**(1):121-125

[67] Zabihollah A, Zareie S. Optimal design of adaptive laminated beam using layerwise finite element. Journal of Sensors. 2011;**2011**:8. Article ID 240341. Available from: https://doi.org/10.1155/2011/240341

[68] Zareie S, Zabihollah A. A failure control method for smart composite morphing airfoil by piezoelectric actuator. Transactions of the Canadian Society for Mechanical Engineering. 2011;**35**(3)

[69] Manoharan R, Vasudevan R, Jeevanantham AK. Dynamic characterization of a laminated composite magnetorheological fluid sandwich plate. Smart Materials and Structures. 2014;**23**(2):25022

[70] Lim SH, Prusty BG, Pearce G, Kelly D, Thomson RS. Study of magnetorheological fluids towards smart energy absorption of composite structures for crashworthiness. Mechanics of Advanced Materials and Structures. 2016;**23**(5):538-544

[71] Rajamohan V, Rakheja S, Sedaghati R. Vibration analysis of a partially treated multi-layer beam with magnetorheological fluid. Journal of Sound and Vibration. 2010;**329**(17):3451-3469

[72] Zabihollah A, Ghafari AS, Yadegari A, Rashidi D. Effects of MR-fluid on low-velocity impact response of MR-laminated beams. In: AIP Conference Proceedings. Vol. 1858, No. 1; 2017. p. 40004

[73] Naji J, Zabihollah A, Behzad M. Vibration characteristics of laminated composite beams with magnetorheological layer using layerwise theory. Mechanics of Advanced Materials and Structures. 2018;**25**(3):202-211. DOI: 10.1080/15376494.2016.1255819

[74] Lee H-G, Sang-Hyun K-WM, Chung L, Lee S-K, Lee M-K, Hwang J-S, et al. Pneumatic motion control using magnetorheological technology. Journal of Intelligent Material Systems and Structures. 2007;**18**(19):1111-1120

Chapter 9

Hysteresis Behavior of Pre-Strained Shape Memory Alloy Wires Subject to Cyclic Loadings: An Experimental Investigation

Shahin Zareie and Abolghassem Zabihollah

Abstract

Shape memory alloys (SMAs) are a class of smart materials with the ability to recover their initial shape after releasing the applied load and experiencing a relatively large amount of strain. However, sequential loading and unloading which is an unavoidable issue in many applications significantly reduces the strain recovery of SMA wires. In the present work, experimental tests have been performed to study the pre-strain effect of SMA wires on hysteresis behavior of SMA under cyclic loadings. The effects of cyclic loading on austenite and martensite properties have been investigated. SMA wires with diameter of 1.5 mm and length of 560 mm subjected to about 1000 cycles show about 3 mm residual deformation, which is approximately equal to 0.5% residual strain. It is observed that applying 1.7% pre-strain on the SMA wire fully eliminates the residual strain due to cyclic loading.

Keywords: shape memory alloy, cyclic loading, pre-strained, residual deformation

1. Introduction

Stability and failure control of structures are of the major challenges for civil structures in which the loading conditions are cyclic in nature, including harbor and offshore structures. In the past decades, a variety of active control mechanisms have been developed to ensure the stability of such structures [1]. However, active control mechanisms are relatively complex, requiring expensive and professional maintenance which are not always accessible. Therefore, for many structures, passive stability control mechanisms are preferred. Shape memory alloys (SMAs), due to their unique ability to recover their initial shape after releasing the applied load, are promising materials for energy absorption and ensuring the stability of structure under relatively large strain [2–7]. However, it is understood that the cyclic loading may highly influence the strain recovery property, superelasticity, of SMAs [8]. This phenomenon has been widely investigated by researchers in smart structure communities.

Zhang et al. [9] developed a low cycle fatigue criterion for superelasticity of SMAs considering thermomechanical coupling. They concluded that increasing strain rate decreases the fatigue lifetime in SMAs. A micromechanical model to describe the cyclic deformation of polycrystalline NiTi under different thermomechanical conditions has been proposed by [10]. Wang et al. [11] proposed a novel structural building element integrated with SMA material to show the capability of SMA-based element in re-centering and energy dissipation of the structure after severe displacement.

Chemisky et al. [12] proposed an analytical model to describe the effects of cyclic loading at high temperature on SMAs behavior. Recently, remarkable researches have been performed to study the different parameters related to cyclic loadings on SMAs' behavior. Kan et al. [13] studied the effect of strain rate on uniaxial deformation of NiTi. Soul et al. [14] investigated the effect of loading frequency on the damping capacity of NiTi wires and concluded that loading frequency above 0.01 Hz significantly reduces the damping capacity of SMA wires. Des Roches et al. [15] evaluated the change in superelasticity of bar and wires made of NiTi under cyclic loading.

The present work aims to investigate the effects of pre-strained SMA wires under cyclic loading conditions on force-displacement behavior, particularly austenite and martensite phases, of SMAs. Experimental tests have been conducted to study the effect of the number of cycles, frequency of loading, and the pre-straining on hysteresis behavior of SMA wires under cyclic loadings.

2. Modeling superelasticity effect of SMA

When the temperature is equal to or greater than the austenitic finish temperature ($T \geq A_f$), SMA materials exhibit the superelasticity effect. **Figure 1** shows the stress-strain curve for loading and unloading of an SMA material. In the loading phase, there are two parts: linear and plastic, whereas in the unloading process, the material exhibits three portions, huge stress reduction with negligible strain reduction, huge strain reduction with a small amount of stress, and finally linear stress-strain reduction. It is noted that the area under the stress-strain curve describes the energy dissipation property of the SMA when loading and unloading. For an SMA rod with cross section (A) subjected to the axial loading (F), the stress

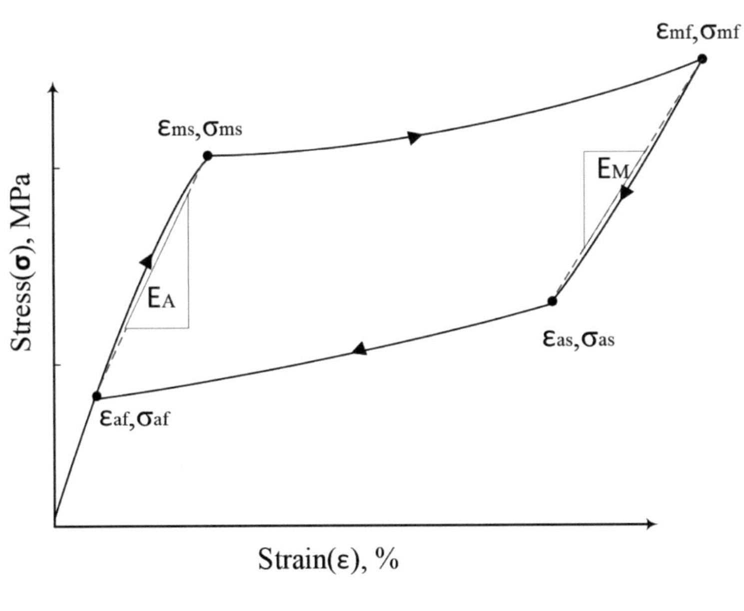

Figure 1.
The schematic diagram of the superelasticity in SMA.

is given by $\sigma = F/A$. Neglecting thermal expansion, the stress-strain of the SMA rod is computed as [16, 17]:

$$\sigma - \sigma_0 = E(\varepsilon - \varepsilon_0) + \Omega(\xi - \xi_0) \tag{1}$$

where $0 \le \xi \le 1$ indicates martensite fraction which is equal to zero for fully austenite phase and 1 for fully martensite. The term Ω is the transformation coefficient. Considering initial state as $\sigma_0 = \varepsilon_0 = 0$, $\xi_0 = 0$ and noting that after full phase transformation the material returns to zero stress, the transformation coefficient may be defined as a function of residual strain, $\Omega = E\varepsilon_r$. Therefore, the stress-strain relations for SMA can be expressed as:

$$\sigma - \sigma_0 = E(\varepsilon - \varepsilon_0) + \varepsilon_r E(\xi - \xi_0) \tag{2}$$

Eq. (2) can be modified for each region, for example, in the linear elastic region where the material is in full austenite phase ($\xi = 0$); Eq. (2) is given as:

$$\sigma = \varepsilon_r \varepsilon E \tag{3}$$

where E_A indicates the modulus of elasticity in austenite phase; similarly, for linear unloading stage where the material is in full martensite phase, modulus of elasticity is defined by E_M. For other regions, modulus of elasticity is a combination of E_A and E_M as the following [16, 17]:

$$E = E_A + (E_M - E_A)\xi \tag{4}$$

Further unloading beyond this point produces a linear elastic behavior to zero stress-strain. For further description on stress-strain relationships for superelasticity effect, one may refer to the book written by [16]. However, one may note that in the above expression, the effect of the number of loading/unloading cycles and pre-straining is not taken into consideration. The following sections provide a thorough study, particularly experimentally, on the effects of cyclic loading and pre-straining on stress-strain behavior of SMA materials.

3. Experimental tests

Experimental tests have been conducted using a universal testing machine—MTS model 370.5 with 500 kN loading capacity, 150 mm dynamic stroke, and 0.1–1 Hz loading frequency as shown in **Figure 2**. The SMA wire specimen of 560 mm length with 1.5 mm diameter made of NiTi has been subjected to cyclic tensile loadings. The material properties of NiTi wire, manufactured by Confluent Medical Technologies Company, are given in **Table 1**. Due to the small diameter of the specimen, each plate is clamped by the top and bottom grippers of the MTS loading machine, as shown in **Figure 3**. Loading frequencies, initial tensile load, loading rates, and the number of cycles have been predefined in the MTS machine for each test.

3.1 Single cycle loading and unloading test

In order to determine the force-displacement response curve for SMA specimen, a NiTi wire described above has been tested at room temperature with a quasi-static loading/unloading, with the period of 20 s and maximum amplitude 12 mm, as

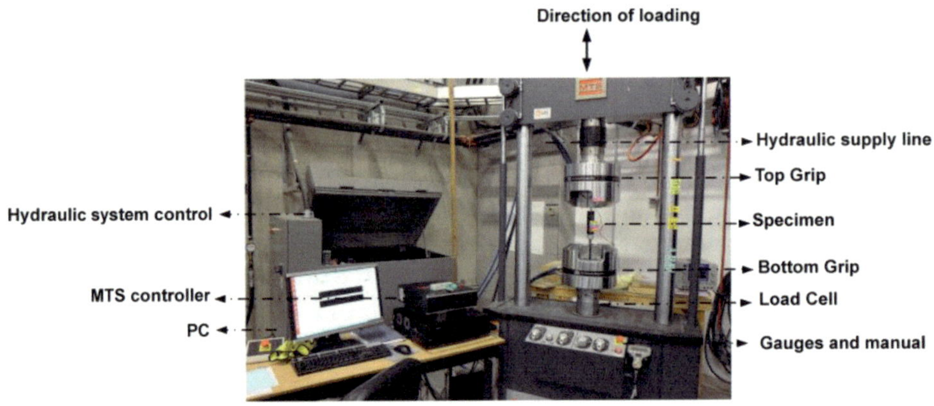

Figure 2.
The MTS loading frame machine and its accessories.

Physical properties	Value
Melting point (°C)	1310
Density (g/cm^3)	6.5
Electrical resistivity (μ ohm-cm)	82
Modulus of elasticity (GPa)	41–75
Coefficient of thermal expansion (/°C)	11×10^{-6}
Ultimate tensile strength (MPa)	~1070
Total elongation	~10%
Straight length (mm)	560
Diameter (mm)	1.5

Table 1.
The properties of NiTi shape memory alloy.

Figure 3.
The experimental setup for griping the SMA specimen.

presented in **Figure 4**. Then, the test specimen is subjected to a cyclic load with 1000 cycles in which the period and maximum amplitude for the first cycle is 1.4 s and 20 mm, correspondingly, as shown in **Figure 5**. After completing 1000 cycles, the specimen is subjected again to the quasi-static load given in **Figure 4**. The force-displacement behavior of the specimen under the first quasi-static load and after

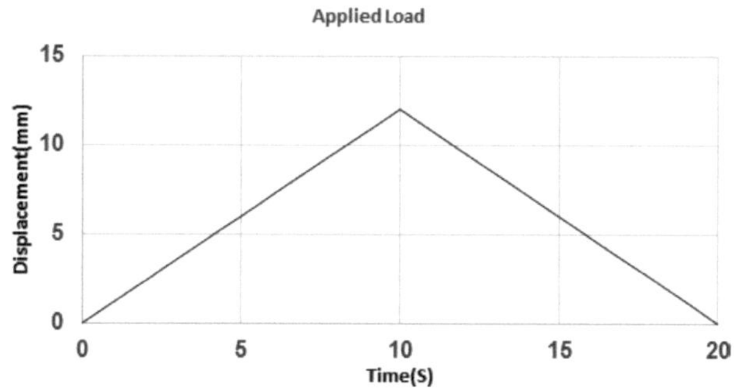

Figure 4.
The quasi-static loading/unloading on the SMA specimen by MTS machine.

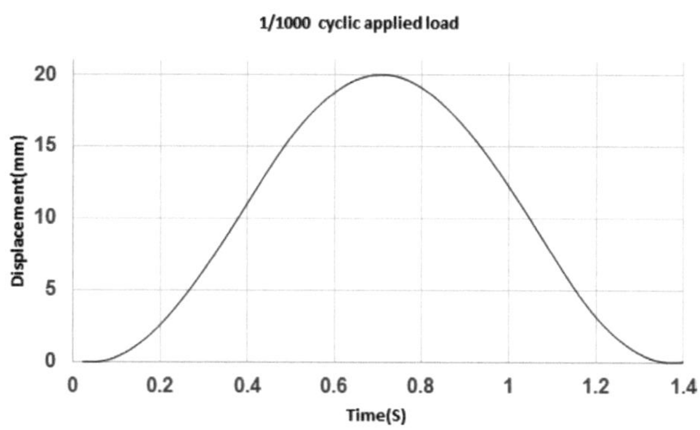

Figure 5.
The first loading cycle on SMA specimen in cyclic loading.

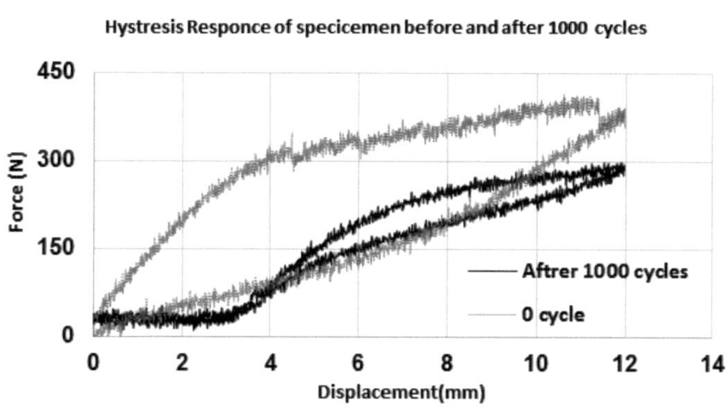

Figure 6.
The effect of cyclic loading on force-displacement.

1000 load cycles are shown in **Figure 6**, in which 0 cycle stands for the quasi-static loading described in **Figure 4**. After 1000 load cycles, two major impacts on the force-displacement curve of SMA are realized, a significant reduction on hysteresis which, in turn, results in reduction in energy absorption, and reduction in material properties, E_A and E_M. Reduction in material properties is mainly due to the degradation phenomena of SMA. The effect of cyclic loading/unloading on material properties, E_A and E_M, are presented as a correction factor of the ratio of E after 1000 cycle (E_{1000}) and E when the specimen is subjected to a single quasi-static load (E_0), in **Figure 6**. After 1000 cycles, E_A shows 18% reduction, whereas E_M 33% compared to the initial single quasi-static loading condition.

3.2 Effect of pre-straining on SMA properties

Long-term stability and performance of civil structures under dynamic loadings is an essential feature for many applications. Therefore, structural designers do not tolerate the change in structural strength and stability. As it was realized in the previous section, cyclic load results in a significant reduction in material properties and, in turn, a reduction in the stability and loading characteristics of SMA-based structural elements.

In order to mitigate the negative effect of cyclic loading on SMA-based elements, 10 mm initial displacement is applied to the SMA wire, which is far beyond the elastic displacement region, and 10 mm is approximately equal to 1.7% pre-strain. The hysteresis response of 1.7% pre-strained SMA wire subjected to the loading protocol, as presented in **Figure 5**, after 1000 cycles, is given in **Figure 7**. Comparison of **Figure 7** and **Figure 6** exhibits that pre-straining SMA eliminates the residual deformation occurred after 1000 cyclic load/unloading.

Figure 8 provides the comparison between E_A and E_M of the specimen after applying 1000 cyclic loads. It is noted that pre-straining improves the value of E_A from 82% at 0% pre-strain to 52% at 1.7% pre-straining the specimen. However, 1.7% pre-straining completely return the value of E_M from 67% at 0% to its initial state.

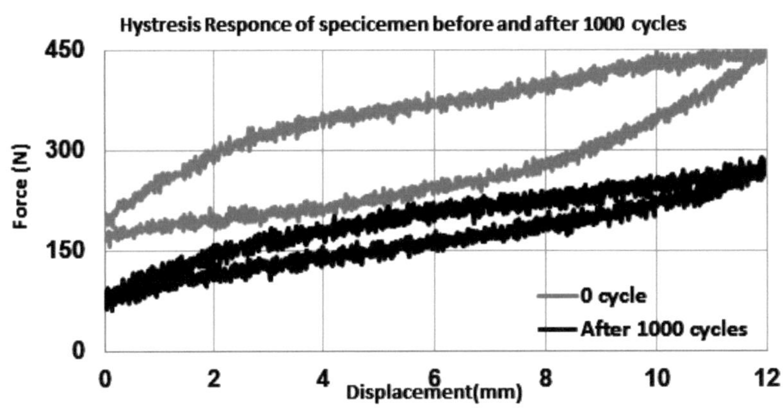

Figure 7.
The hysteresis response of 1.7% pre-strain specimen after applied 1000 cycles.

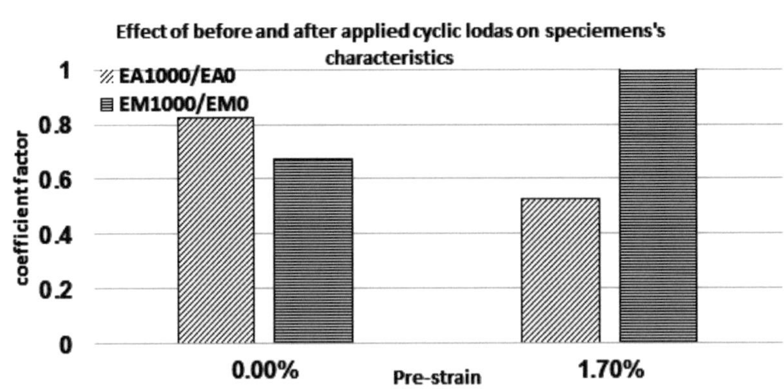

Figure 8.
The effect of applied cyclic loading on characteristics of 0 and 1.7% pre-strain specimen.

3.3 Effect of pre-straining and sequentially increasing load on SMA properties

Sequentially increasing loadings is a common loading profile for many civil applications including offshore structures, requiring a thorough understanding of the response of structures under such loading profile. In this section, two specimens are exposed to a sequentially increasing quasi-static loading protocol as presented in **Figure 9**. It composes of four cycles with 20 s period and arbitrary amplitude of 11.61 mm, (called loading sequence (LS) 1), 13.58 mm (called loading sequence (LS) 2), 17.51 mm (called loading sequence (LS) 3), and 19.48 mm (called loading sequence (LS) 4), respectively. Two parameters, namely, E_A and E_M, have been studied again under this loading protocol. Then, the loading profile is repeated for 1000 cycles, and the response of the specimens is recorded. Similar to the constant loading profile, it is observed that cyclic load leads to a significant reduction in material properties, E_A and E_M. In another experiment, the specimens are subjected to 1.7% pre-straining where it revealed a significant improvement in the material properties. As shown in **Figure 10**, applying a value of 1.7%, pre-strain changes the value of the correction factor E_{A1000}/E_{A0} from 62% (for the first period) to 81% (for the fourth period) to 52 and 72%, correspondingly. In a similar observation, applying a value of 1.7%, pre-straining changes the value of the correction factor E_{M1000}/E_{M0} from 67 to 54%.

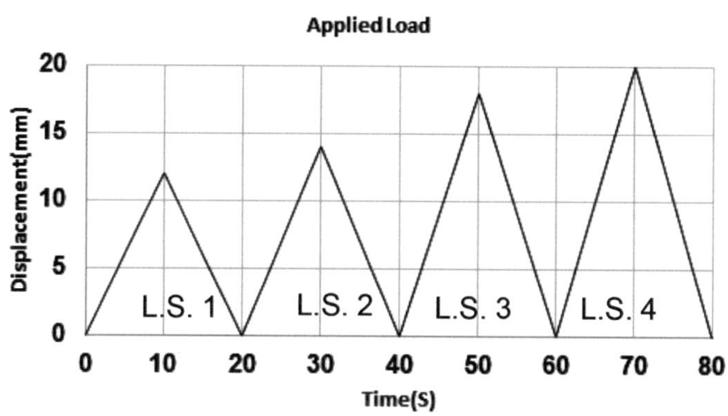

Figure 9.
The loading protocol for 0 and 1.7% pre-strain specimen.

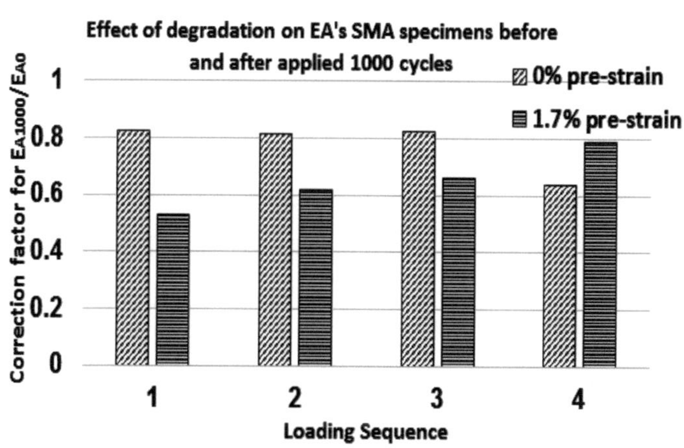

Figure 10.
The effect of cyclic loading on E_A of 0 and 1.7% pre-strain specimen.

4. Correction factor for stress-strain of SMA under cyclic loadings

Close study of force-displacement curves for SMA specimen subjected to cyclic loading, and pre-straining reveals that the number of cycles and pre-straining significantly influences the predetermined stress-strain behavior of SMA-based structural elements. In many applications the structure undergoes dynamic loadings, requiring an accurate yet practical estimation of the effect of loading conditions, on the performance and functionality of the structure for long-term usage. According to the present experimental results, practical correction factors are introduced to estimate the effects of cyclic. According to the present experimental results, it is realized that the change in E_M and E_A is a function of correction factors relating cyclic loading, pre-straining, and initial material properties as the following:

$$E_A^{''} = f(\alpha, \beta, E_A), E_M^{''} = f(\alpha', \beta', E_M) \tag{5}$$

where α and α′ are the correction factors relating E_{A1000}/E_{A0} and E_{M1000}/E_{M0}. Similarly β and β′ are correction factors relating E_{A1000} at 1.7% pre-straining /E_{A1000} at 0% pre-straining and E_{M1000} at 1.7% pre-straining/ E_{M1000} at 0% pre-straining. The terms $E_A^{''}$ and $E_M^{''}$ indicate corrected value for E_A and E_M as a result of cyclic loading and pre-straining. Accordingly, the functions relating $E_A^{''}$ to E_A and $E_M^{''}$ to E_M can be estimated as:

$$E_A^{''} = \alpha\beta E_A, E_M^{''} = \alpha'\beta' E_M \tag{6}$$

The correction factors, $\alpha\beta$ and $\alpha'\beta'$ are presented in **Figures 10** and **11**. The values for α and α′ are given in **Figure 12**.

As observed in the conventional specimen (0% pre-straining), β and β′ are assumed 1. Hence, Eq. (6) is now modified to:

$$E_A^{''} = \alpha E_A, E_M^{''} = \alpha' E_M \tag{7}$$

where $\alpha \approx 0.6 - 0.81$ and $\alpha' \approx 0.6 - 1.2$.

In the case of applied 1.7% pre-straining, coefficients in Eq. (6) can be calculated by $\alpha\beta \approx 0.45–0.8$ and $\alpha'\beta' \approx 0.56 - 0.98$.

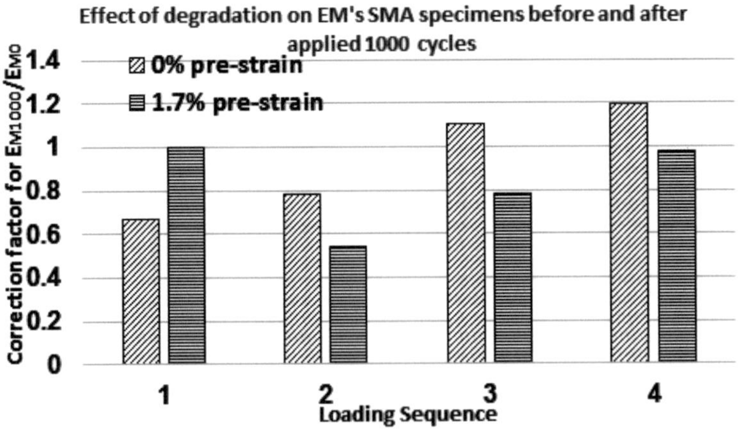

Figure 11.
The effect of cyclic loading on E_M of 0 and 1.7% pre-strain specimen.

Considering the effects of cyclic loading and pre-straining on stress-strain behavior of SMA wires, **Figure 13(a)** is now modified to **Figure 13(b)**, where one may realize a significant reduction in critical stress-strain points as well as changes in residual strain after cyclic loading.

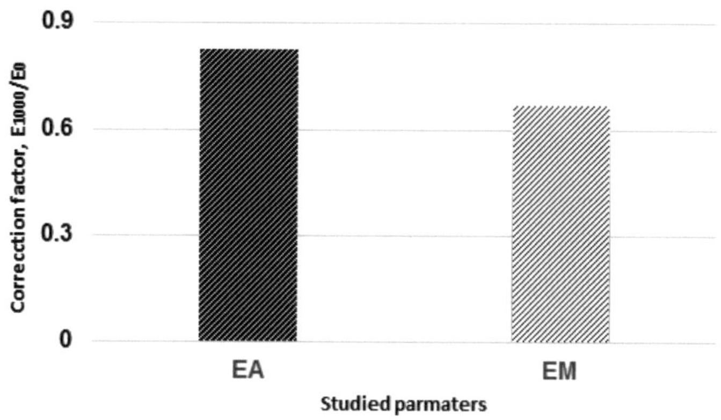

Figure 12.
The effect of applied cyclic loading on EA and EM.

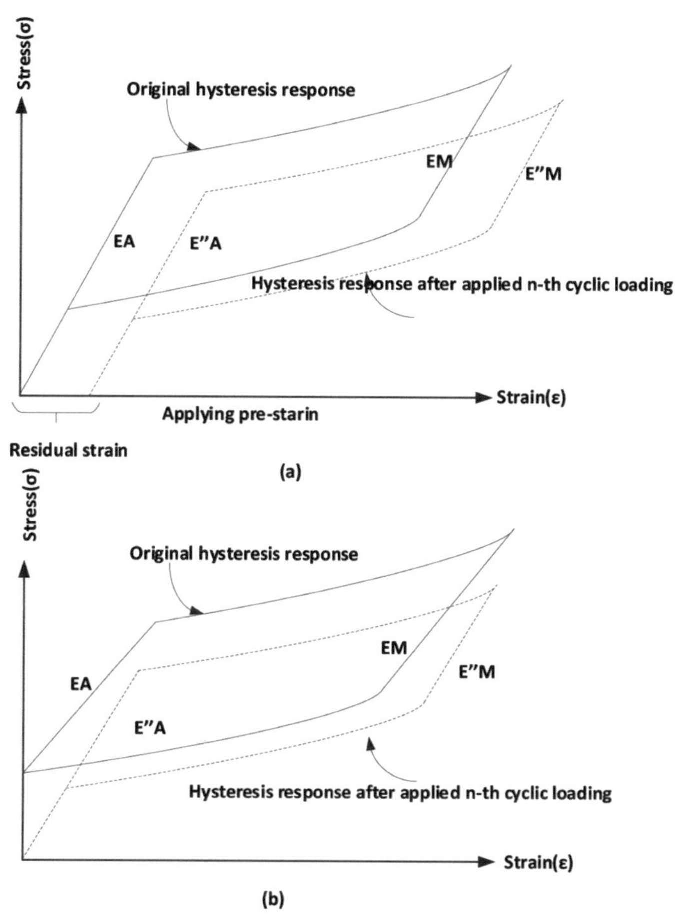

Figure 13.
The effect of cyclic loading (a) 0% pre-strained SMA (b) 1.7% pre-strained SMA on EA and EM.

5. Conclusions

The effects of cyclic loading and pre-straining on stress-strain behavior of SMA wires have been investigated. Several experimental tests have been conducted to study the effects of the number of cycles and pre-straining on strain recovery and modulus of elasticity of SMA wires. Correction factors have been provided to encounter the effects of cyclic loads and pre-straining on stress-strain behavior of SMA wires. It is observed that cyclic loading may lead to huge residual strain and reduces the strain recovery feature of SMA. However, pre-straining SMA for only 1.7% significantly improves the reduction in strain recovery of SMA as a result of cyclic loading.

Acknowledgements

This research received funding from the Green Construction Research and Training Center (GCRTC). The authors wish to thank the University of British Columbia-Okanagan Campus, for the technical support. The authors also wish to thank Mr. Kyle Charles as a research engineer and ex lab manager for his valuable help in experimental tests at the applied laboratory for advanced materials and structures at the University of British Columbia-Okanagan Campus.

Author details

Shahin Zareie[1]* and Abolghassem Zabihollah[2]

1 School of Engineering, The University of British Columbia, Kelowna, BC, Canada

2 School of Science and Engineering, Sharif University of Technology, International Campus, Kish Island, Iran

*Address all correspondence to: shahin@alumni.ubc.ca

References

[1] Aryan H, Ghassemieh M. A superelastic protective technique for mitigating the effects of vertical and horizontal seismic excitations on highway bridges. Journal of Intelligent Material Systems and Structures. 2017;**28**(12):1533-1552

[2] Zareie S, Mirzai N, Alam MS, Seethlaer RJ. A dynamic analysis of a novel shape memory alloy-based bracing system. In: CSCE 2017; 2017

[3] Zareie S, Mirzai N, Alam MS, Seethlaer RJ. An introduction and modeling of novel shape memory alloy-based bracing. In: CSCE 2017; 2017

[4] Zareie S, Alam MS, Seethaler RJ, Zabihollah A. An experimental study of SMA wire for tendons of a tension leg platform. In: 5th Annual Engineering Graduate Symposium; 2019

[5] Zareie S, Shahria Alam M, Seethlaer RJ, Zabihollah A. Effect of shape memory alloy-magnetorheological fluid-based structural control system on the marine structure using nonlinear time-history analysis. Journal of Applied Ocean Research. 2019;**91**:101836

[6] Zareie S, Shahria Alam M, Seethlaer RJ, Zabihollah A. An experimental study of SMA wire for tendons of a tension leg platform. In: 5th Annual Engineering Graduate Symposium; 2019

[7] Zareie S, Alam MS, Seethaler RJ, Zabihollah A. Effect of cyclic loads on SMA-based component of cable-stayed bridge. In: 7th International Specialty Conference on Engineering Mechanics and Materials; 2019

[8] Miyazaki S, Imai T, Igo Y, Otsuka K. Effect of cyclic deformation on the pseudoelasticity characteristics of Ti-Ni alloys. Metallurgical Transactions A. 1986;**17**(1):115-120

[9] Zhang Y, Zhu J, Moumni Z, Van Herpen A, Zhang W. Energy-based fatigue model for shape memory alloys including thermomechanical coupling. Smart Materials and Structures. 2016;**25**(3):35042

[10] Yu C, Kang G, Kan Q, Song D. A micromechanical constitutive model based on crystal plasticity for thermo-mechanical cyclic deformation of NiTi shape memory alloys. International Journal of Plasticity. 2013;**44**:161-191

[11] Wang W, Chan T-M, Shao H, Chen Y. Cyclic behavior of connections equipped with NiTi shape memory alloy and steel tendons between H-shaped beam to CHS column. Engineering Structures. 2015;**88**:37-50

[12] Chemisky Y, Chatzigeorgiou G, Kumar P, Lagoudas DC. A constitutive model for cyclic actuation of high-temperature shape memory alloys. Mechanics of Materials. 2014;**68**:120-136

[13] Kan Q, Yu C, Kang G, Li J, Yan W. Experimental observations on rate-dependent cyclic deformation of super-elastic NiTi shape memory alloy. Mechanics of Materials. 2016;**97**:48-58

[14] Soul H, Isalgue A, Yawny A, Torra V, Lovey FC. Pseudoelastic fatigue of NiTi wires: Frequency and size effects on damping capacity. Smart Materials and Structures. 2010;**19**(8):85006

[15] DesRoches R, McCormick J, Delemont M. Cyclic properties of superelastic shape memory alloy wires and bars. Journal of Structural Engineering. 2004;**130**(1):38-46

[16] Leo DJ. Engineering Analysis of Smart Material Systems. John Wiley & Sons; 2007

[17] Zuo X-B, Li A-Q, Sun W, Sun X-H. Optimal design of shape memory alloy damper for cable vibration control. Journal of Vibration and Control. 2009;**15**(6):897-921

Chapter 10

Interactional Modeling and Optimized PD Impedance Control Design for Robust Safe Fingertip Grasping

Izzat Al-Darraji, Ali Kılıç and Sadettin Kapucu

Abstract

Dynamic and robust control of fingertip grasping is essential in robotic hand manipulation. This study introduces detailed kinematic and dynamic mathematical modeling of a two-fingered robotic hand, which can easily be extended to a multi-fingered robotic hand and its control. The Lagrangian technique is applied as a common procedure to obtain the complete nonlinear dynamic model. Fingertip grasping is considered in developing the detailed model. The computed torque controller of six proportional derivative (PD) controllers, which use the rotation angle of the joint variables, is proposed to linearize the hand model and to establish trajectory tracking. An impedance controller of optimized gains is suggested in the control loop to regulate the impedance of the robotic hand during the interaction with the grasped object in order to provide safe grasping. In this impedance controller, the gains are designed by applying a genetic algorithm to reach minimum contact position and velocity errors. The robustness against disturbances is achieved within the overall control loop. A computer program using MATLAB is used to simulate, monitor, and test the interactional model and the designed controllers.

Keywords: modeling robotic hand, optimized impedance controller, linearization the dynamics of robotic hand, fingertip grasping, Lyapunov stability, Lagrange technique

1. Introduction

A recent research on service robotics has shown that there is an environmental reaction on major applications [1]. The robotic hand is the main interactive part of the environment with the service robotics [2, 3]. Grasping operation [4] has been used explicitly in robotic hand during the interactive period which represents the process of holding objects via two stages: free motion and constrained motion. Free motion is the moving of fingers before detecting the grasped object, and the constrained motion is the state of fingers when the object is detected. The common purpose in grasping is to handle unknown objects. Thus, finding both *a model system* and *a control algorithm* to deal with non-predefined objects is essential in a robotic hand. Regarding modeling, interactional modeling is the demand to perform a specific control algorithm. With interactional modeling, the contact force between the robotic hand and the grasped objects has obtained without needing their properties. In control point of view, the robotic hand holds an object by its fingers and moves the object to another place. In this case, there are different modes of operation which can be listed as before holding the grasped object, holding the grasped object, and moving the robotic hand while holding the grasped object. The aim at each mode is distinct to the others and has to respond to the operation assigned. Thus, the multitude of various goals which have to be achieved with regard to the various operations presents an interesting scenario in the control of a robotic hand.

Generally, it is essential to obtain the modeling of a multi-fingered robotic hand before implementing a specific model-based control algorithm. Arimoto et al. [5] presented a two-fingertip grasping model to control the rolling contact between the fingertips and the grasped object. Arimoto et al. derived the model through the use of the Euler-Lagrange equation with respect to the geometry of the fingertips which is represented by the arc length of a fingertip. Corrales et al. [6] developed a model for a three-fingered robotic hand; in their study the transmission of fingertip contact force from the hand to the grasped object is considered. Boughdiri et al. [7] discussed the issue of modeling a multi-fingered robotic hand as a first step, before designing controllers. They considered free motion only, without the constraints of grasped objects. In a proceeding work, Boughdiri et al. later [8] considered the constraints of grasped objects and the coupling between fingers in grasping tasks.

In the control scenario of a robotic-environment interaction, there is generally also a similar case of a robotic hand-grasped object interaction; force tracking is necessary to obtain appropriate and safe contacts. In force tracking, three main position-force controllers are introduced: stiffness control, hybrid position/force control, and impedance control [9]. Among the above three controllers, impedance control presents an integrated process to constrained manipulators [10]. The impedance controller is introduced by Hogan in 1985 to set the dynamic conduct between the motion of the robot end-effector and the contact force with the environment, instead of regarding the control of motion and force individually [11, 12]. Chen et al. [13] derived impedance controller for an elastic joint-type robotic hand in Cartesian and joint spaces. Chen et al. focused on the compensation of gravity and managing the parameter of uncertainties. Huang et al. [14] proposed the impedance controller for grasping by two fingers to deal with the problem of dropping objects when grasping by one finger. Zhang et al. [15] proposed an adaptive impedance controller with friction compensation to track forces of the multi-fingered robotic hand.

However, the synthesis of impedance control design continues to exist as a major challenge in the performance advancement of robotic hand in order to provide safe and robust grasping. Besides this, the greatest problems in robotic hand-grasped object interaction are the existence of nonlinearity in the dynamic system, dealing with different grasped objects, disturbances, and interference. In the previous studies, the aforementioned problems have not taken in consideration. The aim of this paper is to treat these problems using interactional modeling and the design gains of impedance control. The action of the suggested control expression is to provide a fingertip interactional robust control algorithm that takes into account the dynamics of the robotic hand under unknown contact conditions. To accomplish this, we propose first to linearize the dynamics on the level of position-tracking controller through six PD controllers. When executed, the calculated gains of the designed PD controllers result in zero steady errors without overshoot. Second, through an impedance controller, the interactional fingertips' force is regulated by the robotic hand's impedance. The gains of the impedance are designed based on genetic algorithm. As a result, the introduced control expression enables the position and velocity of fingertips to be controlled simultaneously without changing over control subspace. This control expression avoids selecting between different controllers which causes indirect control and consumption of high energy. The operation and the robustness of the presented control expression are approved by MATLAB/Simulink in several tests.

The rest of the chapter is organized as follows. In Section 2, the mathematical representation of kinematics and dynamics of the robotic hand with environment interaction is derived and explained. In Section 3, the overall controllers of position tracking and impedance control are designed. In addition, the suggested procedure of obtaining optimum values of impedance gains using genetic algorithm is explained. In Section 4, following the joint path and fingertip position is tested. Then, the robustness of the system is examined. Finally, the conclusions of this study are presented in Section 5.

2. Mathematical model

The models of a robotic hand system are the mathematical equations that describe all the contact forces and motion of the fingers at any time [16]. These models are essential for grasping and manipulation operations which include kinematics, dynamics, and contacts between the fingers and the environment. A multi-fingered robotic hand with an existing grasped object has a complex structure. Thus, deriving the mathematical representation is not a simple task. In this section, the mathematical representation for a two-fingered robotic hand with an environment is developed. Changing the orientation of the robotic hand with respect to the inertial (earth) frame during manipulation operation is considered in this analysis.

2.1 Reference frames

The system is a two-fingered robotic hand that can adjust fingertip grasp forces by the torque of actuators in the joints. Each finger includes three joints (MCP, PIP, and DIP) and three rigid phalanges (proximal, intermediate, and distal). The motion of the finger actuates through the torque in the joints. The variables $i = 1, 2$ and $j = 1, 2, 3$ represent the index of the finger and the joint, respectively. The analysis is based on the four main reference frames: finger frames O_{ij}, hand base frame O_{hb}, fingertip contact point frames O_{tip_i}, and inertia frame O_e as shown in **Figure 1**. The joints MCP, PIP, and DIP are located at O_{i1}, O_{i2}, and O_{i3}, respectively. The length L_{ij} represents the distance from O_{ij} to $O_{i(j+1)}$. The angle of rotation θ_{ij} around z_{j-1} associates with the ith phalanx and j^{th} joint. The coordinate frame O_{ij} is attached in order to express the coordinates of any point on the phalanges as a result of the motion of θ_{ij}. In our case, expressing the motion of the fingertip points $(x_{tip_1}, y_{tip_1}, z_{tip_1})$ and $(x_{tip_2}, y_{tip_2}, z_{tip_2})$ is the primary interest. The frame O_{hb} is defined to switch between the motions of the fingertips with respect to the base of the hand through matrices. The frames O_{tip_i} are defined to represent the contact points between the fingertips and the environment. Finally, the inertial frame O_e which is assumed at O_{hb} is assigned to represent the orientation of the robotic hand with respect to the earth.

The result of orienting the robotic hand with respect to the earth coordinates is expressed by defining the three rotation angles roll, pitch, and yaw (**RPY**). In which, roll is (α: around x_{hb}-axis), pitch is (γ: around y_{hb}-axis), and yaw is (φ: around z_{hb}-axis). Mathematical representation of rotation about a fixed axis (earth frame) can be expressed by a specified sequence of multiplying the free

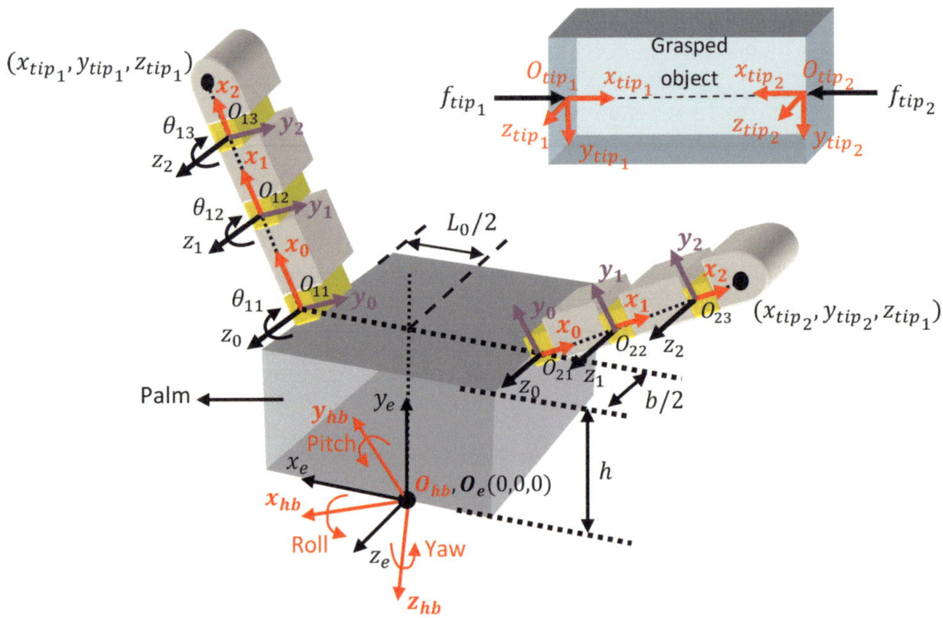

Figure 1.
Coordinate frames of the robotic hand.

rotation matrices Rx(α), Ry(γ), and Rz(φ) [17]. Thus, the final rotation matrix R_e^{hb} which switches from the earth frame $O_e(x_e, y_e, z_e)$ to the robotic hand base frame is obtained as

$$\begin{aligned} R_e^{hb} &= R(z_e y_e x_e, yaw, pitch, roll) \\ &= R(z_e, \varphi) R(y_e, \gamma) R(x_e, \alpha) \\ &= \begin{bmatrix} c_\varphi c_\gamma & c_\varphi s_\gamma s_\alpha - s_\varphi c_\alpha & c_\varphi s_\gamma c_\alpha + s_\varphi s_\alpha \\ s_\varphi c_\gamma & s_\varphi s_\gamma s_\alpha + c_\varphi c_\alpha & s_\varphi s_\gamma c_\alpha - c_\varphi s_\alpha \\ -s_\gamma & c_\gamma s_\alpha & c_\gamma c_\alpha \end{bmatrix}. \end{aligned} \quad (1)$$

where s and c indicate the symbols of the sine and cosine functions, respectively.

2.2 Kinematic analysis

Forward kinematics is implemented here to express the relationship between the Cartesian spaces of the fingertip contact points, which are represented by coordinates $(x_{tip_i}, y_{tip_i}, z_{tip_i})$ relative to the joint position θ_{ij}. The positions of fingertip 1 and fingertip 2 change by rotating the joints in clockwise and counterclockwise directions, respectively. According to the coordinate frames and the dimension parameters of **Figure 1**, the forward kinematic of finger 1 is expressed as

$$T_{tip_1}^{hb} = \begin{bmatrix} c_{123} & -s_{123} & 0 & \frac{L_0}{2} + L_1 c_1 + L_2 c_{12} + L_3 c_{123} \\ s_{123} & c_{123} & 0 & h + L_1 s_1 + L_2 s_{12} + L_3 s_{123} \\ 0 & 0 & 1 & -\frac{b}{2} \\ 0 & 0 & 0 & 1 \end{bmatrix}, \quad (2)$$

and the forward kinematic of finger 2 is

$$T^{hb}_{tip_2} = \begin{bmatrix} c_{123} & -s_{123} & 0 & -\frac{L_0}{2} - L_1c_1 - L_2c_{12} - L_3c_{123} \\ s_{123} & c_{123} & 0 & h + L_1s_1 + L_2s_{12} + L_3s_{123} \\ 0 & 0 & 1 & -\frac{b}{2} \\ 0 & 0 & 0 & 1 \end{bmatrix}. \tag{3}$$

where L_0 and b are the length and the width of the palm shown in **Figure 1**, respectively. In Eqs. (2) and (3) and the rest of equations in this study, the denotation $c_1, s_1, c_{12}, s_{12}, c_{123}$, and s_{123} represents $\cos\theta_1$, $\sin\theta_1$, $\cos(\theta_1+\theta_2)$, $\sin(\theta_1+\theta_2)$, $\cos(\theta_1+\theta_2+\theta_3)$, and $\sin(\theta_1+\theta_2+\theta_2)$, respectively. To control the position of the three links, the joint variables $\theta_{i1}, \theta_{i2}, \theta_{i3}$ are calculated using inverse kinematic as

$$\theta_{i2} = \pi - \cos^{-1}\frac{L_1{}^2 + L_2{}^2 - x_{tip_i}{}^2 - y_{tip_i}{}^2}{2L_1L_2}, \tag{4}$$

$$\theta_{i1} = \tan^{-1}\frac{y_{tip_i}}{x_{tip_i}} - \cos^{-1}\frac{x_{tip_i}{}^2 + y_{tip_i}{}^2 + L_1{}^2 - L_2{}^2}{2l_1\sqrt{x_{tip_i}{}^2 + y_{tip_i}{}^2}}, \tag{5}$$

$$\theta_{i3} = \varphi_{ie} - \theta_1 - \theta_2. \tag{6}$$

where φ_{ie} is the angle of distal phalanx with respect to the axis y_{i2}. Next, the Jacobian matrix for each finger is obtained based on the forward kinematic equations, which were derived from Eqs. (2) and (3) as

$$J_{finger_1} = \begin{bmatrix} -L_1s_1 - L_2s_{12} - L_3s_{123} & -L_2s_{12} - L_3s_{123} & -L_3s_{123} \\ L_1c_1 + L_2c_{12} + L_3c_{123} & L_2c_{12} + L_3c_{123} & L_3c_{123} \\ 0 & 0 & 0 \end{bmatrix}, \tag{7}$$

and

$$J_{finger_2} = \begin{bmatrix} L_1s_1 + L_2s_{12} + L_3s_{123} & L_2s_{12} + L_3s_{123} & L_3s_{123} \\ L_1c_1 + L_2c_{12} + L_3c_{123} & L_2c_{12} + L_3c_{123} & L_3c_{123} \\ 0 & 0 & 0 \end{bmatrix}, \tag{8}$$

The robotic hand transforms the generated torque by each joint to the wrenches exerted on the grasped object through the following Jacobian matrix:

$$J_{hand} = \begin{bmatrix} J_{finger_1} & 0 \\ 0 & J_{finger_2} \end{bmatrix}. \tag{9}$$

2.3 Contact interaction

A contact interaction, which uses contact behavior between rigid fingertips and a flexible grasped object, is proposed in this study (see **Figure 2**). The grasped object is assumed as a linear spring of stiffness k_{object}. Hence, each fingertip exerts a force on the grasped object as

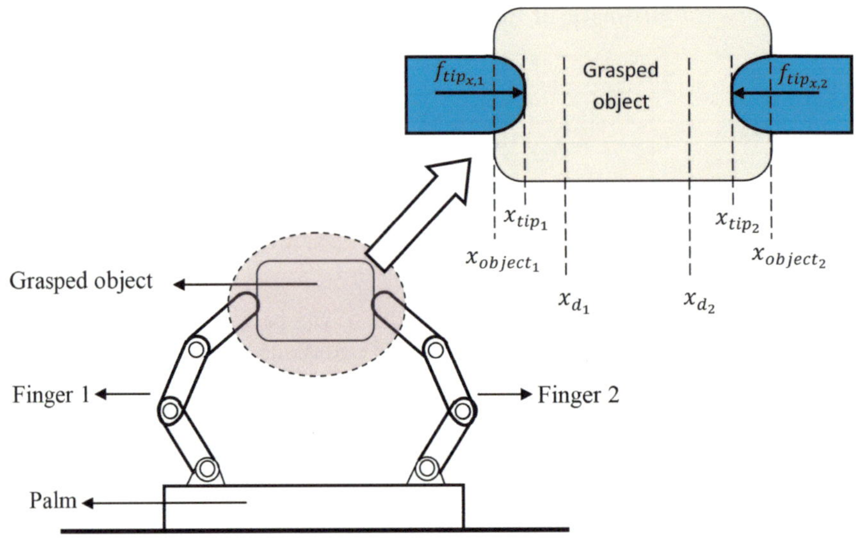

Figure 2.
Side view of fingertip contact interaction system on the x-axis.

$$f_{tip_{x,i}} = k_{object}\left(x_{tip_{x,i}} - x_{object_{x,i}}\right), \tag{10}$$

$$f_{tip_{y,i}} = k_{object}\left(y_{tip_{y,i}} - y_{object_{y,i}}\right), \tag{11}$$

$$f_{tip_{z,i}} = k_{object}\left(z_{tip_{z,i}} - z_{object_{z,i}}\right). \tag{12}$$

where $x_{object_i}, y_{object_i}$, and z_{object_i} are the coordinate points of the grasped object. The current and the desired location of the fingertips are denoted by $(x_{tip_i}, y_{tip_i}, z_{tip_i})$ and $(x_{d_i}, y_{d_i}, z_{d_i})$, respectively.

2.4 Dynamics of robotic hand

The motion equations are derived to describe the dynamics of the robotic hand when interacting with the grasped object; the explained fingertip forces in Eqs. (10)–(12) are involved in the dynamic equation. When ignoring friction, each finger in free motion can be described by the following general motion equation:

$$M_{finger_1}(q_1)\ddot{q}_1 + D_{finger_1}(q_1, \dot{q}_1)\dot{q}_1 + G_{finger_1}(q_1) = \tau_{finger_1}, \tag{13}$$

$$M_{finger_2}(q_2)\ddot{q}_2 + D_{finger_2}(q_2, \dot{q}_2)\dot{q}_2 + G_{finger_2}(q_2) = \tau_{finger_2}, \tag{14}$$

where $q_i \in R^3$ is the general coordinate vector, $M_{finger_i}(q_i)$ is the 3×3 inertia matrix, $D_{finger_i}(q_i, \dot{q}_i)$ is the 3-vector which involves forces of centrifugal and Coriolis, $G_{finger_i}(q_i)$ is the vector of the gravitational force, and τ_{finger_i} is the vector of the input torque at joints. When the fingertips are interacting with the grasped object (see **Figure 2**), then Eqs. (13) and (14) become

$$M_{finger_1}(q_1)\ddot{q}_1 + D_{finger_1}(q_1, \dot{q}_1)\dot{q}_1 + G_{finger_1}(q_1) = \tau_{finger_1} - J^T_{finger_1} F_{tip_1}, \tag{15}$$

$$M_{finger_2}(q_2)\ddot{q}_2 + D_{finger_2}(q_2, \dot{q}_2)\dot{q}_2 + G_{finger_2}(q_2) = \tau_{finger_2} - J^T_{finger_2} F_{tip_2}, \tag{16}$$

where J_{finger_i} is the Jacobian matrix (see Subsection 2.2) and F_{tip_i} is the external force 3-vector with expression value:

$$F_{tip_1} = \left[f_{tip_{x,1}}\ f_{tip_{y,1}}\ f_{tip_{z,1}}\right]^T \text{ and } F_{tip_2} = \left[f_{tip_{x,2}}\ \ f_{tip_{y,2}}\ \ f_{tip_{z,2}}\right]^T;$$

to derive Eqs. (15) and (16), the Lagrange equations of motion are used. For each finger manipulator, the equation of motion is given by [18]

$$\frac{d}{dt}\frac{\partial T}{\partial \dot{\theta}_j} - \frac{\partial T}{\partial \theta_j} = Q_j + T_j. \tag{17}$$

where T is the total kinetic energy of the finger manipulator, Q_j is the potential energy, and T_j is the external torques. The x-y coordinates are represented according to the assigned frames of **Figure 1**. The velocities of the mass centers for the phalanges are important to find the kinetic energy. For proximal phalanx, the position of the mass center with respect to MP joint is represented by

$$\begin{aligned}\overline{x}_1 &= d_1 c_1,\\ \overline{y}_1 &= d_1 s_1;\end{aligned}$$

the position of the mass center of intermediate phalanx with respect to MP joint is represented by

$$\begin{aligned}\overline{x}_2 &= L_1 c_1 + d_2 c_{12}\\ \overline{y}_2 &= L_1 s_1 + d_2 s_{12};\end{aligned}$$

and the position of the mass center of distal phalanx with respect to MP joint is represented by

$$\begin{aligned}\overline{x}_3 &= L_1 c_1 + L_2 c_{12} + d_3 c_{123},\\ \overline{y}_3 &= L_1 s_1 + l_2 s_{12} + d_3 s_{123},\end{aligned}$$

where d_i is the distance to the center of mass which is assumed in the middle of the phalanx. The total kinetic energy of the finger in Eq. (17) is expressed by the following equation:

$$\begin{aligned}T =& \frac{1}{2}m_1\left(\dot{\overline{x}}_1^{\,2} + \dot{\overline{y}}_1^{\,2}\right) + \frac{1}{2}I_1\dot{\theta}_1^{\,2} + \frac{1}{2}m_2\left(\dot{\overline{x}}_2^{\,2} + \dot{\overline{y}}_2^{\,2}\right) + \frac{1}{2}I_2\left(\dot{\theta}_1 + \dot{\theta}_2\right)^2\\ &+ \frac{1}{2}m_3\left(\dot{\overline{x}}_3^{\,2} + \dot{\overline{y}}_3^{\,2}\right) + \frac{1}{2}I_3\left(\dot{\theta}_1 + \dot{\theta}_2 + \dot{\theta}_3\right)^2,\end{aligned} \tag{18}$$

and to obtain the kinetic energy, the above formulas $(\overline{x}_1, \overline{y}_1, \overline{x}_2, \overline{y}_2, \overline{x}_3, \overline{y}_3)$ are differentiated with time. Lagrangian equations according to Eq. (17) are

$$\frac{d}{dt}\frac{\partial T}{\partial \dot{\theta}_1} - \frac{\partial T}{\partial \theta_1} = Q_1 + T_1, \tag{19}$$

$$\frac{d}{dt}\frac{\partial T}{\partial \dot{\theta}_2} - \frac{\partial T}{\partial \theta_2} = Q_2 + T_2, \tag{20}$$

$$\frac{d}{dt}\frac{\partial T}{\partial \dot{\theta}_3} - \frac{\partial T}{\partial \theta_3} = Q_3 + T_3, \tag{21}$$

where

$$Q_1 = F_1 \frac{\partial \vec{r}_1}{\partial \theta_1} + F_2 \frac{\partial \vec{r}_2}{\partial \theta_1} + F_3 \frac{\partial \vec{r}_3}{\partial \theta_1}, \tag{22}$$

$$Q_2 = F_1 \frac{\partial \vec{r}_1}{\partial \theta_2} + F_2 \frac{\partial \vec{r}_2}{\partial \theta_2} + F_3 \frac{\partial \vec{r}_3}{\partial \theta_2}, \tag{23}$$

$$Q_3 = F_1 \frac{\partial \vec{r}_1}{\partial \theta_3} + F_2 \frac{\partial \vec{r}_2}{\partial \theta_3} + F_3 \frac{\partial \vec{r}_3}{\partial \theta_3}. \tag{24}$$

The procedure for calculating Q_i is shown in Appendix A. Thus, Eqs. (22)–(24) become

$$\begin{aligned} Q_1 = {} & -m_1 g d_1 \cos\theta_1 - m_2 g l_1 \cos\theta_1 - m_2 g d_2 \cos(\theta_1 + \theta_2) - m_3 g l_1 \cos\theta_1 \\ & -m_3 g l_2 \cos(\theta_1 + \theta_2) - m_3 g d_3 \cos(\theta_1 + \theta_2 + \theta_3), \end{aligned} \tag{25}$$

$$Q_2 = -m_2 g d_2 \cos(\theta_1 + \theta_2) - m_3 g l_2 \cos(\theta_1 + \theta_2) m_3 g d_3 \cos(\theta_1 + \theta_2 + \theta_3), \tag{26}$$

$$Q_3 = -m_3 g d_3 \cos(\theta_1 + \theta_2 + \theta_3); \tag{27}$$

by using the kinetic energy in Eq. (18) and the potential energy in Eqs. (25)–(27), it results that Eqs. (19)–(21) after being rearranged in a matrix form result in

$$\begin{bmatrix} m_{11} & m_{12} & m_{13} \\ m_{21} & m_{22} & m_{23} \\ m_{31} & m_{32} & m_{33} \end{bmatrix} \begin{bmatrix} \ddot{\theta}_{i1} \\ \ddot{\theta}_{i2} \\ \ddot{\theta}_{i3} \end{bmatrix} + \begin{bmatrix} d_{11} & d_{12} & d_{13} \\ d_{21} & d_{22} & d_{23} \\ d_{31} & d_{32} & d_{33} \end{bmatrix} \begin{bmatrix} \dot{\theta}_{i1} \\ \dot{\theta}_{i2} \\ \dot{\theta}_{i3} \end{bmatrix} + \begin{bmatrix} g_{i1} \\ g_{i2} \\ g_{i3} \end{bmatrix} = \tau_{finger_i} - J_{finger_i}^T F_{tip_i}. \tag{28}$$

The formula of matrices' elements in Eq. (28) is listed in Appendix A. From Eqs. (15) and (16), the overall dynamics of robotic hand can be constituted as

$$M_{hand}(q_1, q_2) \begin{bmatrix} \ddot{q}_1 \\ \ddot{q}_2 \end{bmatrix} + D_{hand}(q_1, \dot{q}_1, q_2, \dot{q}_2) \begin{bmatrix} \dot{q}_1 \\ \dot{q}_2 \end{bmatrix} + G_{hand}(q_1, q_2) = \tau_{hand} - F_{tips}, \tag{29}$$

with the following matrix forms;

$$M_{hand} = \begin{bmatrix} M_{finger_1}(q_1) & 0 \\ 0 & M_{finger_2}(q_2) \end{bmatrix},$$

$$D_{hand} = \begin{bmatrix} D_{finger_1}(q_1, \dot{q}_1) & 0 \\ 0 & D_{finger_2}(q_2, \dot{q}_2) \end{bmatrix},$$

$$G_{hand} = \begin{bmatrix} G_{finger_1}(q_1) \\ G_{finger_2}(q_2) \end{bmatrix},$$

$$\tau_{hand} = \begin{bmatrix} \tau_{finger_1} \\ \tau_{finger_2} \end{bmatrix},$$

$$F_{tips} = J_{hand}^T \begin{bmatrix} F_{tip_1} \\ F_{tip_2} \end{bmatrix}.$$

Thus, the six motion equations are presented by rearranging Eq. (29) to result in

$$\begin{bmatrix} \ddot{q}_1 \\ \ddot{q}_2 \end{bmatrix} = M_{hand}(q_1, q_2)^{-1}\left(-D_{hand}(q_1, \dot{q}_1, q_2, \dot{q}_2)\begin{bmatrix} \dot{q}_1 \\ \dot{q}_2 \end{bmatrix} - G_{hand}(q_1, q_2) + \tau_{hand} - F_{tips}\right). \quad (30)$$

The desired finger configuration in joint space can be obtained by solving the above equation, which will be explained later in Section 3.1 for designing the position-tracking controller.

3. Controller design

The derived mathematical representation is applied using MATLAB/Simulink to implement the proposed controllers and the tests of this study. The parameters of the model, which are used in Simulink, are shown in **Table 1**. The overall controllers of the robotic hand are depicted as a complete block diagram in **Figure 3**. In the following sections, the design of the controllers is described in detailed.

3.1 Position tracking

Recently, a variant of the PD controller explicitly has been considered in the control of nonlinear system where the aim is to make the system follow a specific trajectory [19–22]. The joint angle controller shown in **Figure 3** is the inner feedforward loop which presents the effect of the input torque on the error in Eq. (30). The detailed block diagram of this controller is shown in **Figure 4**. The tracking error is formed by subtracting the joint angle vector (q_i) from the desired joint angle vector (q_{d_i}) as below:

$$e_i(t) = q_{d_i}(t) - q_i(t). \quad (31)$$

The objective is to minimize this error at any time, by differentiating Eq. (31) twice, solving for $(\ddot{q}_1, \ddot{q}_2)$ in Eq. (30), and rearranging the results in terms of the input torque that yields the following computed torque control equation:

$$\tau_{hand} = M_{hand}\left(\begin{bmatrix} \ddot{q}_{d_1} \\ \ddot{q}_{d_2} \end{bmatrix} - \begin{bmatrix} \ddot{e}_1 \\ \ddot{e}_2 \end{bmatrix}\right) + D_{hand} + G_{hand} + F_{tips}, \quad (32)$$

Parameter	Value	Parameter	Value
L_1	50 mm	L_3	20 mm
d_1	25 mm	d_3	10 mm
m_1	34 g	m_3	12 g
I_1	7083 g.mm^2	I_3	400 g.mm^2
L_2	30 mm	h	100 mm
d_2	15 mm	b	50 mm
m_2	15 g	L_0	70 mm
I_2	1125 g.mm^2		

Table 1.
Robotic hand model parameters.

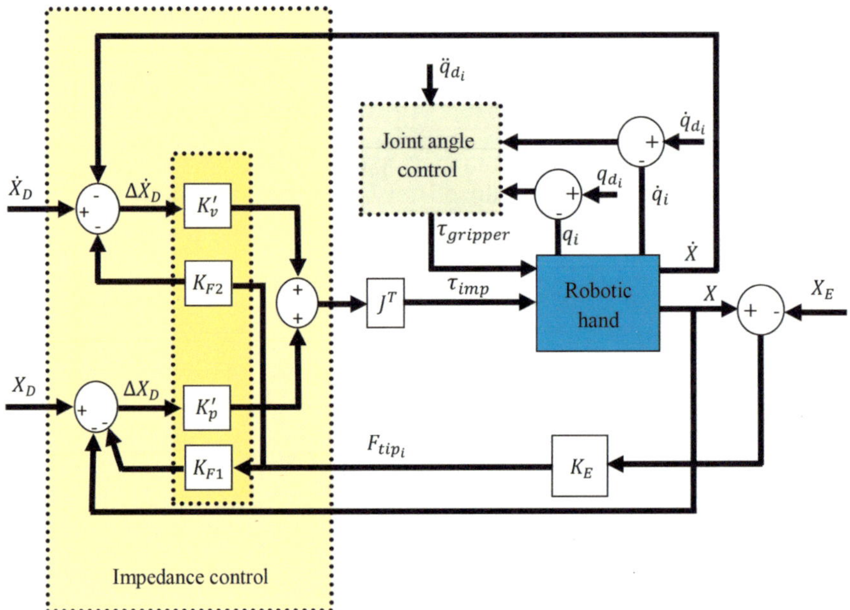

Figure 3.
Block diagram of the complete robotic hand controllers.

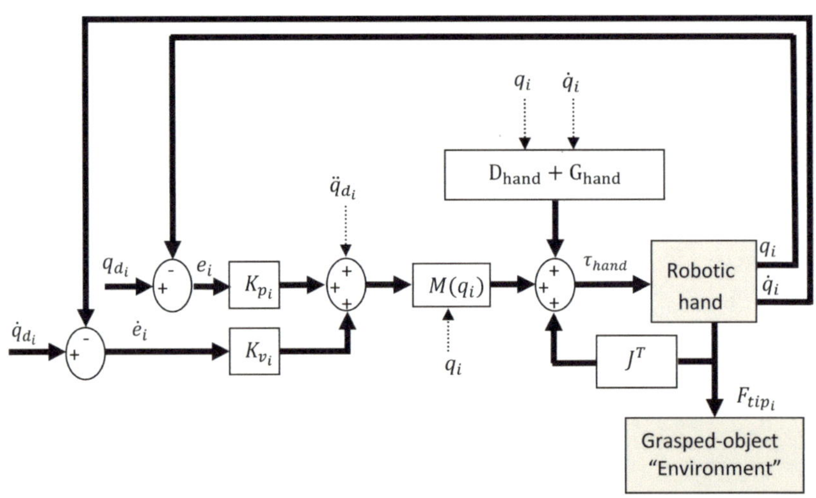

Figure 4.
PD joint angle controller for each finger.

where $\ddot{e}_i$ is the control input which is selected as six PD controllers with the following feedback:

$$\ddot{e}_i = -K_{v_i}\dot{e}_i - K_{p_i}e_i, \tag{33}$$

where $K_{v_i} = \text{diag}\{k_{v_i}\}$ and $K_{p_i} = \text{diag}\{k_{p_i}\}$; hence, the robotic hand input becomes

$$\tau_{hand} = M_{hand}\left(\begin{bmatrix}\ddot{q}_{d_1}\\ \ddot{q}_{d_2}\end{bmatrix} + \begin{bmatrix}K_{v_1}\dot{e_1} + K_{p_1}e_1\\ K_{v_2}\dot{e}_2 + K_{p_2}e_2\end{bmatrix}\right) + D_{hand} + G_{hand} + F_{tips}. \tag{34}$$

The system can be stabilized in Eq. (34) by a proper design of PD controller to track the error $error(t)$ to zero. In which, the input torque at joints results in the trajectory of reaching the grasped object. According to Eq. (33), the typical transfer function of the error dynamic in closed-loop approach is

$$\frac{error(s)}{w(s)} = \frac{1}{s^2 + k_v s + k_p}, \tag{35}$$

where w is the disturbance. Thus, the system is stable for all positive values of k_v and k_p. Comparing Eq. (35) with the standard second-order polynomial term $(s^2 + 2\lambda\omega_n s + \omega_n^2)$, we get

$$k_{p_i} = \omega_n^2,$$
$$k_{v_i} = 2\lambda\omega_n.$$

The design of gains is selected for the closed-loop time constant $(0.1s)$; thus

$$\omega_n = \frac{1}{0.1} = 10,$$
$$k_p = \omega_n{}^2 = 100,$$
$$k_v = 2\omega_n = 20$$

where ω_n and λ are the natural frequency and the damping coefficient, respectively.

The stability of robustness analysis by Lyapunov's direct method is implemented here to verify the ability of applying the proposed PD controller in real situations where unmodeled disturbances are included in the actual robotic hand. The effect of unmodeled disturbances in robotic manipulator can be represented by replacing the PD controller gain matrices with matrices of nonlinear gains [23]. Considering the nonlinear matrices gains, the robotic hand input in Eq. (34) becomes

$$\tau_{hand} = M_{hand}\left(\begin{bmatrix}\ddot{q}_{d_1}\\ \ddot{q}_{d_2}\end{bmatrix} + \begin{bmatrix}K_{v_1}(e_1)\dot{e_1} + K_{p_1}(e_1)e_1\\ K_{v_2}(e_2)\dot{e}_2 + K_{p_2}(e_2)e_2\end{bmatrix}\right) + D_{hand} + G_{hand} + F_{tips}, \tag{36}$$

where $K_{p_i}(e_i) = \mathrm{diag}\{k_{p_i}(e_i)\}$ and $K_{v_i}(e_i) = \mathrm{diag}\{k_{v_i}(e_i)\}$ are the nonlinear proportional gain matrix and the nonlinear derivative gain matrix, respectively. Combining Eq. (32) with Eq. (36), the equation of the closed loop becomes

$$\frac{d}{dt}\begin{bmatrix}e_1\\ e_2\\ \dot{e_1}\\ \dot{e}_2\end{bmatrix} = \begin{bmatrix}\dot{e_1}\\ \dot{e}_2\\ -K_{v_1}(e_1)\dot{e_1} - K_{p_1}(e_1)e_1\\ -K_{v_2}(e_2)\dot{e}_2 - K_{p_2}(e_2)e_2\end{bmatrix}. \tag{37}$$

The stability of Eq. (32) together with Eq. (36) is assumed if there exists $\rho > 0$ with the following inequality sentences:

$$k_{p_i}(e_i) \geq \rho,$$
$$k_{v_i}(e_i) \geq \rho.$$

The stability analysis is implemented by proposing the following Lyapunov function:

$$V = \frac{1}{2}\dot{e}_1{}^T\dot{e}_1 + \frac{1}{2}\dot{e}_2{}^T\dot{e}_2 + \int_0^{e_1} \boldsymbol{\varepsilon}_1{}^T\left[K_{p_1}(\boldsymbol{\varepsilon}_1) + \beta_1 K_{v_1}(\boldsymbol{\varepsilon}_1)\right]d\boldsymbol{\varepsilon}_1 + \int_0^{e_2} \boldsymbol{\varepsilon}_2{}^T\left[K_{p_2}(\boldsymbol{\varepsilon}_2) + \beta_2 K_{v_2}(\boldsymbol{\varepsilon}_2)\right]d\boldsymbol{\varepsilon}_2$$
$$+ \beta_1 e_1{}^T\dot{e}_1 + \beta_2 e_2{}^T\dot{e}_2 \,. \tag{38}$$

Considering the smallest eigenvalues λ_{si}, the limit of β_i satisfies

$$\lambda_{s1}\{K_{v_1}(\boldsymbol{\varepsilon}_1)\} > \beta_1 > 0,$$
$$\lambda_{s2}\{K_{v_2}(\boldsymbol{\varepsilon}_2)\} > \beta_2 > 0;$$

thus

$$k_{v_{j1}}(\varepsilon_{j1}) - \frac{\beta_1}{2} > 0,$$
$$k_{v_{j2}}(\varepsilon_{j2}) - \frac{\beta_2}{2} > 0,$$

where $j = 1, \ldots, 3$. Since $\beta_1, \beta_2 > 0$ and $k_{p_{j1}}(\varepsilon_{j1}), k_{p_{j2}}(\varepsilon_{j2}) \geq \rho$; thus

$$k_{p_{j1}}(\varepsilon_{j1}) + \beta_1 k_{v_{j1}}(\varepsilon_{j1}) - \frac{\beta_1{}^2}{2} \geq \rho, \tag{39}$$

$$k_{p_{j2}}(\varepsilon_{j2}) + \beta_2 k_{v_{j2}}(\varepsilon_{j2}) - \frac{\beta_2{}^2}{2} \geq \rho. \tag{40}$$

Then, Eq. (38) becomes

$$V = \frac{1}{2}(\dot{e}_1 + \beta_1 e_1)^T(\dot{e}_1 + \beta_1 e_1) + \frac{1}{2}(\dot{e}_2 + \beta_2 e_2)^T(\dot{e}_2 + \beta_2 e_2)$$
$$+ \int_0^{e_1} \boldsymbol{\varepsilon}_1{}^T\left[K_{p_1}(\boldsymbol{\varepsilon}_1) + \beta_1 K_{v_1}(\boldsymbol{\varepsilon}_1) - \frac{\beta_1{}^2}{2}I\right]d\boldsymbol{\varepsilon}_1 + \int_0^{e_2} \boldsymbol{\varepsilon}_2{}^T\left[K_{p_2}(\boldsymbol{\varepsilon}_2) + \beta_2 K_{v_2}(\boldsymbol{\varepsilon}_2) - \frac{\beta_2{}^2}{2}I\right]d\boldsymbol{\varepsilon}_2. \tag{41}$$

The first two terms of Lyapunov function are positive. Regarding the third and the fourth terms, according to Eqs. (39) and (40), we have

$$\int_0^{e_{j1}} \varepsilon_1\left[k_{p_{j1}}(\varepsilon_{j1}) + \beta_1 k_{v_{j1}}(\varepsilon_{j1}) - \frac{\beta_1{}^2}{2}\right]d\varepsilon_1 \geq \frac{1}{2}\rho\left|e_{j1}\right|^2, \tag{42}$$

$$\int_0^{e_{j2}} \varepsilon_2\left[k_{p_{j2}}(\varepsilon_{j2}) + \beta_2 k_{v_{j2}}(\varepsilon_{j2}) - \frac{\beta_2{}^2}{2}\right]d\varepsilon_2 \geq \frac{1}{2}\rho\left|e_{j2}\right|^2; \tag{43}$$

thus

$$\int_0^{e_{j1}} \varepsilon_1\left[k_{p_{j1}}(\varepsilon_{j1}) + \beta_1 k_{v_{j1}}(\varepsilon_{j1}) - \frac{\beta_1{}^2}{2}\right]d\varepsilon_1 \to \infty \text{ when } \left|e_{j1}\right| \to \infty, \tag{44}$$

$$\int_0^{e_{j2}} \varepsilon_2\left[k_{p_{j2}}(\varepsilon_{j2}) + \beta_2 k_{v_{j2}}(\varepsilon_{j2}) - \frac{\beta_2{}^2}{2}\right]d\varepsilon_2 \to \infty \text{ when } \left|e_{j2}\right| \to \infty. \tag{45}$$

Equations (44) and (45) result in

$$\int_0^{e_1} \boldsymbol{\varepsilon}_1{}^T\left[K_{p1}(\boldsymbol{\varepsilon}_1) + \beta_1 K_{v1}(\boldsymbol{\varepsilon}_1) - \frac{\beta_1{}^2}{2}\right]d\boldsymbol{\varepsilon}_1 \geq \frac{1}{2}\rho\|e_1\|^2, \tag{46}$$

$$\int_0^{e_2} \boldsymbol{\varepsilon}_2{}^T \left[K_{p_2}(\boldsymbol{\varepsilon}_2) + \beta_2 K_{v2}(\boldsymbol{\varepsilon}_2) - \frac{\beta_2{}^2}{2}\right] d\boldsymbol{\varepsilon}_2 \geq \frac{1}{2}\rho\|e_2\|^2, \tag{47}$$

and

$$\int_0^{e_1} \boldsymbol{\varepsilon}_1{}^T \left[K_{p_1}(\boldsymbol{\varepsilon}_1) + \beta_1 K_{v1}(\boldsymbol{\varepsilon}_1) - \frac{\beta_1{}^2}{2}\right] d\boldsymbol{\varepsilon}_1 \rightarrow \infty \; when \; \|e_1\| \rightarrow \infty, \tag{48}$$

$$\int_0^{e_2} \boldsymbol{\varepsilon}_2{}^T \left[K_{p_2}(\boldsymbol{\varepsilon}_2) + \beta_2 K_{v2}(\boldsymbol{\varepsilon}_2) - \frac{\beta_2{}^2}{2}\right] d\boldsymbol{\varepsilon}_2 \rightarrow \infty \; when \; \|e_2\| \rightarrow \infty. \tag{49}$$

Hence, the Lyapunov function presented in Eq. (38) is globally positive definitive function.

Taking the derivative of Eq. (38) with respect to time, we get

$$\dot{V} = \dot{e}_1{}^T\ddot{e}_1 + \dot{e}_2{}^T\ddot{e}_2 + e_1{}^T\left[K_{p_1}(e_1) + \beta_1 K_{v1}(e_1)\right]\dot{e}_1 + e_2{}^T\left[K_{p_2}(e_2) + \beta_2 K_{v2}(e_2)\right]\dot{e}_2 + \beta_1\dot{e}_1{}^T\dot{e}_1 + \beta_1 e_1{}^T\ddot{e}_1 + \beta_2\dot{e}_2{}^T\dot{e}_2 + \beta_2 e_2{}^T\ddot{e}_2 \tag{50}$$

Finally, using the rule of Leibnitz for representation, the integrals in differentiation form and Substituting Eq. (37) in (50) yield

$$\dot{V} = -\dot{e}_1{}^T[K_{v1}(e_1) - \beta_1 I]\dot{e}_1 - \dot{e}_2{}^T[K_{v2}(e_2) - \beta_2 I]\dot{e}_2 - \beta_1 e_1{}^T K_{p_1}(e_1)e_1 - \beta_2 e_2{}^T K_{p_2}(e_2)e_2, \tag{51}$$

for all positive values of gain matrices $K_{v1}(e_1), K_{v2}(e_2), K_{p_1}(e_1)$,and $K_{p_2}(e_2)$; Eq. (51) is globally negative function. Then

$$\dot{V} < 0.$$

In this way, the system is globally exponential stabile.

3.2 Impedance controller

The PD impedance controller has been widely applied in modern researches for controlling the interaction with the environment [24, 25]. The fundamental principle of the implemented impedance control in this study is that the fingers should track a motion trajectory, and adjust the mechanical impedance of the robotic hand. The mechanical impedance of the robotic hand is defined in terms of velocity and position as

$$Z_{hand} = \frac{F_{tip_i}(s)}{\dot{X}(s)}, \tag{52}$$

$$sZ_{hand} = \frac{F_{tip_i}(s)}{X(s)}. \tag{53}$$

By virtue of the above two equations, the fingertip contact force F_{tip_i} gives the essential property of regulating the position and velocity of the contact points between the fingertips and the grasped object. This regulating behavior is obtained by two PD controllers as shown in **Figure 3**. In the first PD controller, the position is regulated by

$$\Delta X_{stiffness} = K_{F1} F_{tip_i}, \tag{54}$$

where K_{F1} is the matrix of the proportional gain of the controller and the velocity is regulated by

$$\Delta \dot{X}_{damping} = K_{F2} F_{tip_i}, \tag{55}$$

where K_{F2} is the matrix of the controller's derivative gain. The regulated position and velocity of Eqs. (54) and (55) are integrated in the impedance control loop with another PD controller of proportional gain matrix K'_p and derivative gain matrix K'_v in order to adjust the damping of the robotic hand during grasping operation according to the following equation:

$$\tau_{imp} = J^T \left(K'_p \, \Delta X_D + K'_v \Delta \dot{X}_D \right), \tag{56}$$

where

$$\Delta X_D = X_D - X - \Delta X_{stiffness},$$

and

$$\Delta \dot{X}_D = \dot{X}_D - \dot{X} - \Delta \dot{X}_{damping}.$$

Hence, the contact forces are regulated in joint space, and the robotic hand can adapt the collision of the fingers with the grasped objects.

The application of Lyapunov function technique determines the conditions of stability of impedance controller [26]. In the forward kinematic model, the equations of fingers are

$$X_i = f(q_i), \tag{57}$$

$$\dot{X}_i = J_{finger_i} \dot{q}_i. \tag{58}$$

On the other hand, the errors are given as

$$\beth_i = q_i - q_{id}, \tag{59}$$

$$\aleph_i(\beth_i) = X_i(q_i) - X_{id}, \tag{60}$$

where $\beth_i$, $\aleph_i(\beth_i)$, and i are the error of joint angle, the error in task space, and the index of finger ($i = 1, 2$), respectively. Consider the constrained dynamics of the two fingers in Eqs. (15) and (16) for the control torque in Eq. (56). Applying Eqs. (59) and (60), the error dynamic equation of each finger becomes

$$M_{finger_1}(\beth_1)\ddot{\beth}_1 + D_{finger_1}(\beth_1, \dot{\beth}_1)\dot{\beth}_1 + G_{finger_1}(\beth_1) + J^T_{finger_1}\left[K'_p + K'_p K_{F1} K_e + K'_v K_{F2} K_e + K_e\right]\aleph_1(\beth_1) + J^T_{finger_1} K'_v \dot{\aleph}_1(\beth_1) = 0, \tag{61}$$

$$M_{finger_2}(\beth_2)\ddot{\beth}_2 + D_{finger_2}(\beth_2, \dot{\beth}_2)\dot{\beth}_2 + G_{finger_2}(\beth_2) + J^T_{finger_2}\left[K'_p + K'_p K_{F1} K_e + K'_v K_{F2} K_e + K_e\right]\aleph_2(\beth_2) + J^T_{finger_2} K'_v \dot{\aleph}_2(\beth_2) = 0, \tag{62}$$

where K_e is the stiffness matrix of the fingertip/grasped object. In order to find the potential energy $P(\beth_i)$ and the dissipation function $D(\beth_i)$, compare Eqs. (61) and (62) with the following Lagrangian's equations of total kinetic energy T:

$$\frac{d}{dt}\left(\frac{\partial T}{\partial \dot{\beth}_i}\right) - \frac{\partial T}{\partial \beth_i} + \frac{\partial P}{\partial \beth_i} + \frac{\partial D}{\partial \dot{\beth}_i} = 0; \tag{63}$$

then, we get

$$\frac{\partial P}{\partial \beth_1} = \left[K_p' + K_p' K_{F1} K_e + K_v' K_{F2} K_e + K_e\right] \aleph_1(\beth_1), \tag{64}$$

$$\frac{\partial D}{\partial \dot{\beth}_1} = K_v' \dot{\aleph}_1(\beth_1), \tag{65}$$

$$\frac{\partial P}{\partial \beth_2} = \left[K_p' + K_p' K_{F1} K_e + K_v' K_{F2} K_e + K_e\right] \aleph_2(\beth_2), \tag{66}$$

$$\frac{\partial D}{\partial \dot{\beth}_2} = K_v' \dot{\aleph}_2(\beth_2), \tag{67}$$

$$T(\beth_1, \dot{\beth}_1) = \frac{1}{2} \dot{\beth}_1^T M_{finger_1}(\beth_1) \dot{\beth}_1, \tag{68}$$

$$T(\beth_2, \dot{\beth}_2) = \frac{1}{2} \dot{\beth}_2^T M_{finger_2}(\beth_2) \dot{\beth}_2. \tag{69}$$

The error dynamics in Eqs. (61) and (62) is asymptotically stable if the following candidate Lyapunov function

$$V(\beth, \dot{\beth}) = T(\beth_1, \dot{\beth}_1) + T(\beth_2, \dot{\beth}_2) + P(\beth_1) + P(\beth_2) \tag{70}$$

satisfies the following conditions:

$$V(\beth, \dot{\beth}) \text{ is positive,}$$
$$\dot{V}(\beth, \dot{\beth}) \text{ is negative.}$$

The derivative of Eq. (70) is

$$\begin{aligned}
\frac{d}{dt} V(\beth, \dot{\beth}) &= \frac{dT(\beth_1)}{dt} + \frac{dT(\beth_2)}{dt} + \frac{dP(\beth_1)}{dt} + \frac{dP(\beth_2)}{dt}, \\
\dot{V}(\beth, \dot{\beth}) &= \dot{\beth}_1^T M_{finger_1} \ddot{\beth}_1 + \dot{\beth}_1^T \frac{dM_{finger_1}}{dt} \frac{\dot{\beth}_1}{2} + \dot{\beth}_2^T M_{finger_2} \ddot{\beth}_2 + \dot{\beth}_2^T \frac{dM_{finger_2}}{dt} \frac{\dot{\beth}_2}{2} \\
&\quad + \dot{\beth}_1^T \left(\left[K_p' + K_p' K_{F1} K_e + K_v' K_{F2} K_e + K_e\right] \aleph_1(\beth_1)\right) \\
&\quad + \dot{\beth}_2^T \left(\left[K_p' + K_p' K_{F1} K_e + K_v' K_{F2} K_e + K_e\right] \aleph_2(\beth_2)\right).
\end{aligned} \tag{71}$$

Equations (61) and (62) can be rearranged as below

$$\begin{aligned}
&M_{finger_1}(\beth_1)\ddot{\beth}_1 + D_{finger_1}(\beth_1, \dot{\beth}_1)\dot{\beth}_1 + G_{finger_1}(\beth_1) \\
&\quad + J_{finger_1}^T \left[K_p' + K_p' K_{F1} K_e + K_v' K_{F2} K_e + K_e\right] \aleph_1(\beth_1) = -J_{finger_1}^T K_v' \dot{\aleph}_1(\beth_1),
\end{aligned} \tag{72}$$

$$\begin{aligned} &M_{finger_2}(\beth_2)\ddot{\beth}_2 + D_{finger_2}(\beth_2,\dot{\beth}_2)\dot{\beth}_2 + G_{finger_2}(\beth_2) \\ &\quad + J^T_{finger_2}\left[K'_p + K'_p K_{F1} K_e + K'_v K_{F2} K_e + K_e\right]\aleph_2(\beth_2) = -J^T_{finger_2} K'_v \dot{\aleph}_2(\beth_2), \end{aligned} \tag{73}$$

By substituting Eqs. (72) and (73) in Eq. (71) and applying the relations in Eqs. (57)–(60), it yields

$$\dot{V}(\beth,\dot{\beth}) = -\dot{\aleph}_1^T(\beth_1)K'_v\dot{\aleph}_1(\beth_1) - \dot{\aleph}_2^T(\beth_2)K'_v\dot{\aleph}_2(\beth_2). \tag{74}$$

Hence, the robotic hand system is asymptotically stable under the following boundaries:

$$\left[2K'_p + 2K'_p K_{F1} k_e + 2K'_v K_{F2} k_e + 2k_e\right] > 0,$$

$$K'_v > 0,$$

or

$$K'_v > 0.$$

3.3 Gains design using genetic algorithm

Genetic algorithm is a well-known optimization procedure for complex problems which ascertains optimum values for different systems [27]. The genetic algorithm is written to find the optimum impedance gains. The code takes the calculated data from the robotic hand Simulink model to compute impedance's gains. However, the detailed genetic algorithm that is applied in the design of the gains can be defined by the following proceedings:

(1) *Generate a number of solutions*. This represents the random number of population which includes the gains K_{F1}, K_{F2}, K'_p, and K'_v of the impedance controller of **Figure 3**. These gains are the individual of population.

(2) *Fitness function*. During the running of the program, the fitness is evaluated for each individual in the population. In this study, the objective of the fitness function is to minimize the errors ΔX_D and $\Delta\dot{X}_D$ of Eq. (56) as below:

$$\begin{aligned} \mathrm{Error}_{\mathrm{tracking}} = {} & 0.05\Delta X_D + 0.05\Delta\dot{X}_D + 0.1t_{r_1} + 0.1t_{s_1} + 0.1d_{o_1} + 0.1t_{r_2} + 0.1t_{s_2} \\ & + 0.1d_{o_2} + 0.1t_{r_3} + 0.1t_{s_3} + 0.1d_{o_3}, \end{aligned} \tag{75}$$

where t_{r_j}, t_{s_j}, and d_{o_j} are the rising time, the settling time, and the overshoot, respectively, for each error signal ($j = 1, 2, 3$).

(3) *Create the next generation* [28]. At each step, genetic algorithm implements the following three rules to create a new generation of the recent population:

- Individuals called parents are selected by the selection rules, and then the population of the next generation is obtained by contributing the parents.
- Two parents are combined to generate children by crossover rules for the next generation.
- Children are formed by mutation rules. The process of forming children is dependent on applying random changes on the individual parents.

(4) *Setting the genetic algorithm*. The following parameters are used for setting the proposed genetic algorithm in this study: the number of generations = 80, the size of population = 30, and the range of gains are $K_{F1min} = 0$, $K_{F1max} = 50$, $K_{F2min} = 0$, $K_{F2max} = 50$, $K_{P_{min}} = 0$, $K_{P_{max}} = 50$, $K_{Vmin} = 0$, $K_{Vmax} = 50$.

4. Simulation results

The obtained interactional model was applied in the simulation to verify the designed controllers through three tests as follows:

- Test 1: Following a specific joint path.
- Test 2: Following fingertip position.
- Test 3: Robustness of the controllers.

The Simulink model is implemented using MATLAB as shown in **Figure 5**.

4.1 Following a specific joint path

This test is focused on position-tracking performance of joints. We assumed the two fingers are in initial position ($\theta_{ij} = 0$) and the joints should follow the following path:

$$\theta_{ij_d} = 0.2 \sin \pi t.$$

The simulation result explained in **Figure 6** shows that the designed position-tracking controller can follow a specific path with zero steady error and without overshoot. A unit step input function of amplitude 20° is implemented to check this tracking error, and the response of the controllers is more detailed as shown in **Figure 7**.

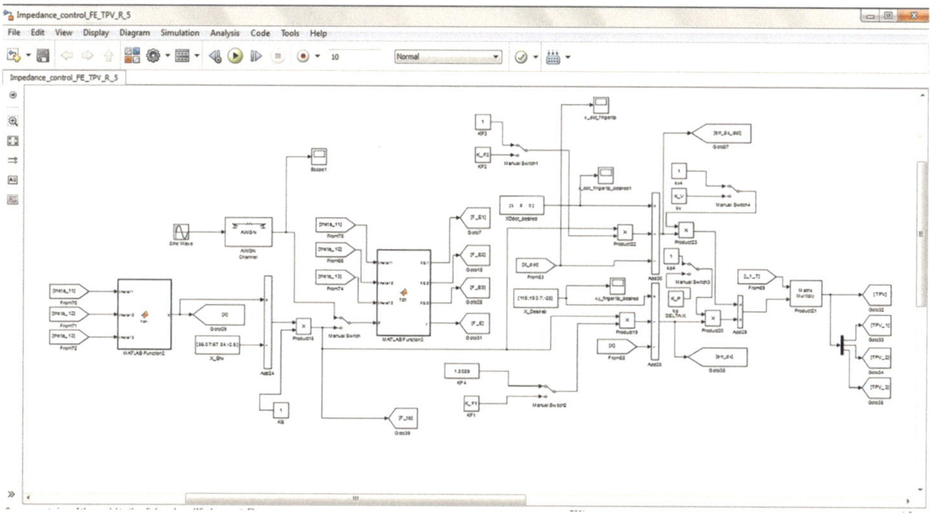

Figure 5.
The Simulink model of the robot hand.

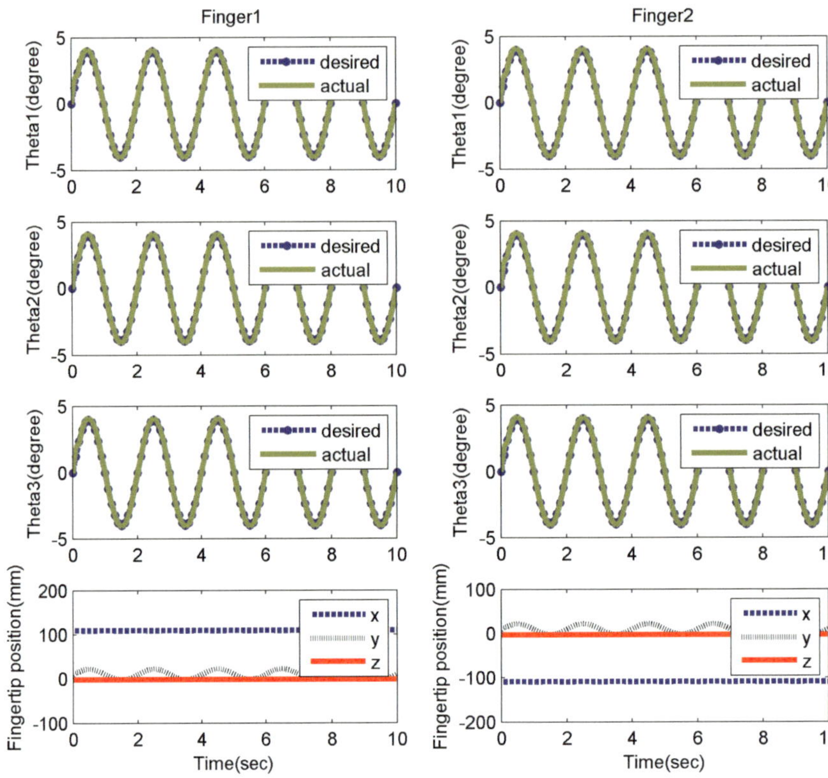

Figure 6.
Joint tracking of sine wave path.

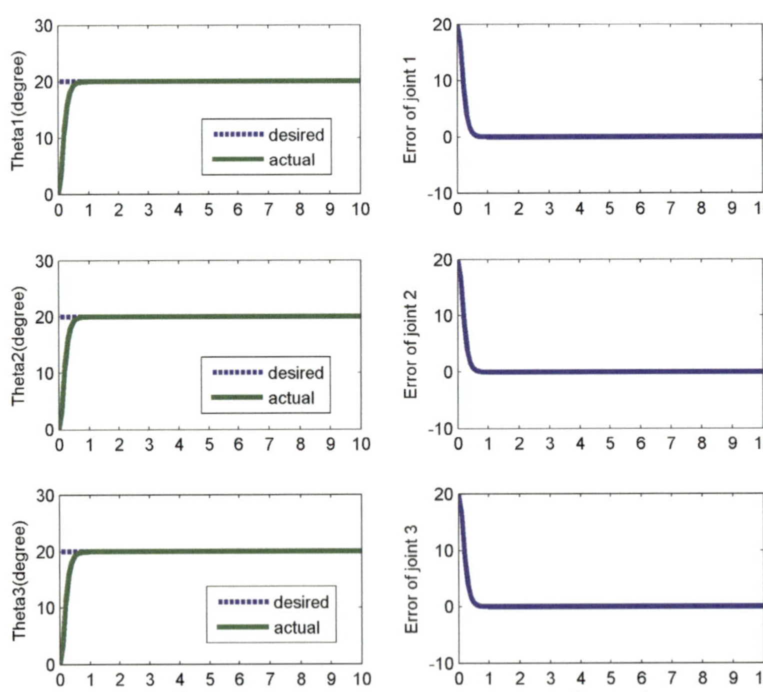

Figure 7.
Tracking errors of finger 1 joints.

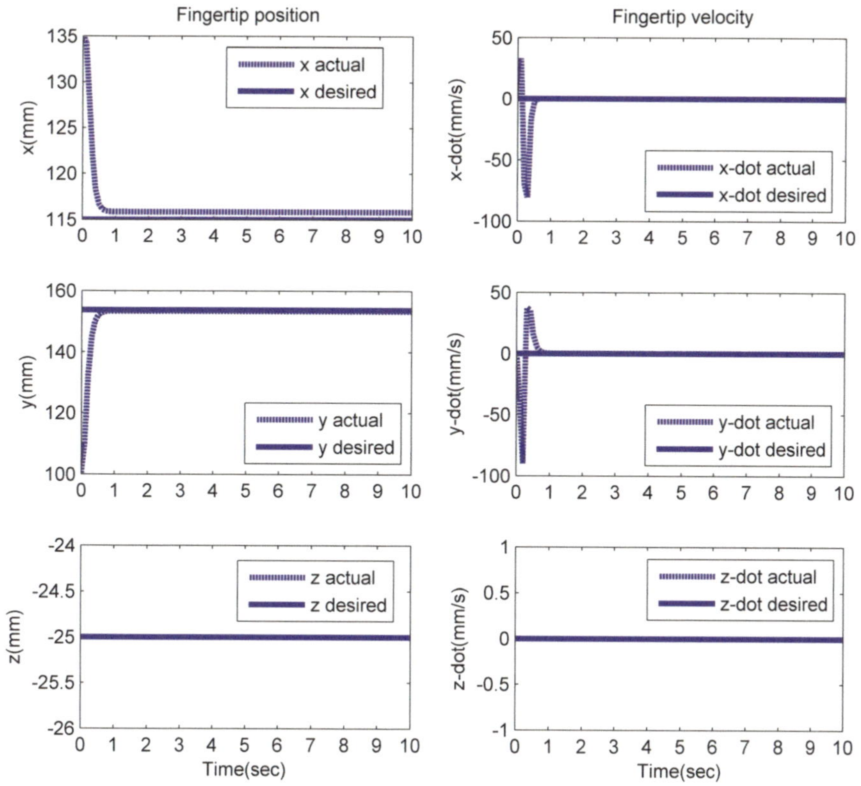

Figure 8.
Motion of fingertip 1 with impedance controller.

4.2 Following fingertip position

The second test is implemented to monitor the operation of the designed impedance controller in task space. We assumed the initial position of the fingertip 1 is (135, 100, −25 mm) and it has to reach the position (115, 153.7, −25 mm). The position of the grasped object is assumed at $x_{object_1} = 118\text{mm}$, $y_{object_1} = 152.7\text{mm}$, $z_{object_1} = -25$. The result of the optimization process of the impedance controller is obtained as $K_{F1} = 1.8089$, $K_{F2} = 1$, $K_P = 1$, $K_V = 1$. At these gains, as shown in **Figure 8**, the percentage error of position and velocity are 0.6 and 0%, respectively. Regarding the fingertip contact force represented in **Figure 9**, the overall designed controllers have given the essential property of contact without overshoot.

For comparison purposes, another study is implemented without considering the impedance controller in the overall control loop, i.e., $K_{F1} = 0$, $K_{F2} = 0$, $K_P = 0$, $K_V = 0$. The results showed that the fingertip position error reaches 0.69%, but the main side effect was of the fingertip velocity error which reached 100% (see **Figure 10**). In turn, the fingertip contacts the grasped object with high value of error in velocity; this situation causes unsafe grasping. As a result, the proposed overall controller has minimized the velocity error to 0% thanks to the proposed impedance controller with genetic algorithm.

4.3 System robustness

The robustness of the designed controllers is checked in this section. This test is implemented within the same parameters of the desired position and object location

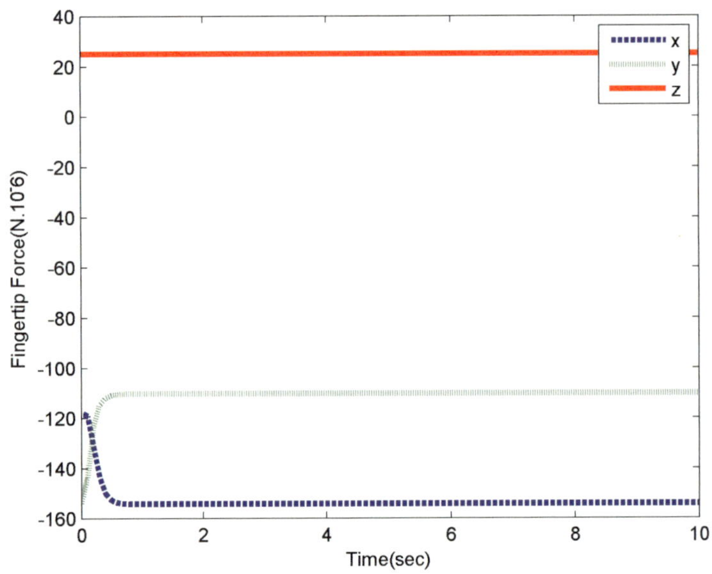

Figure 9.
Fingertip contact force with the grasped object.

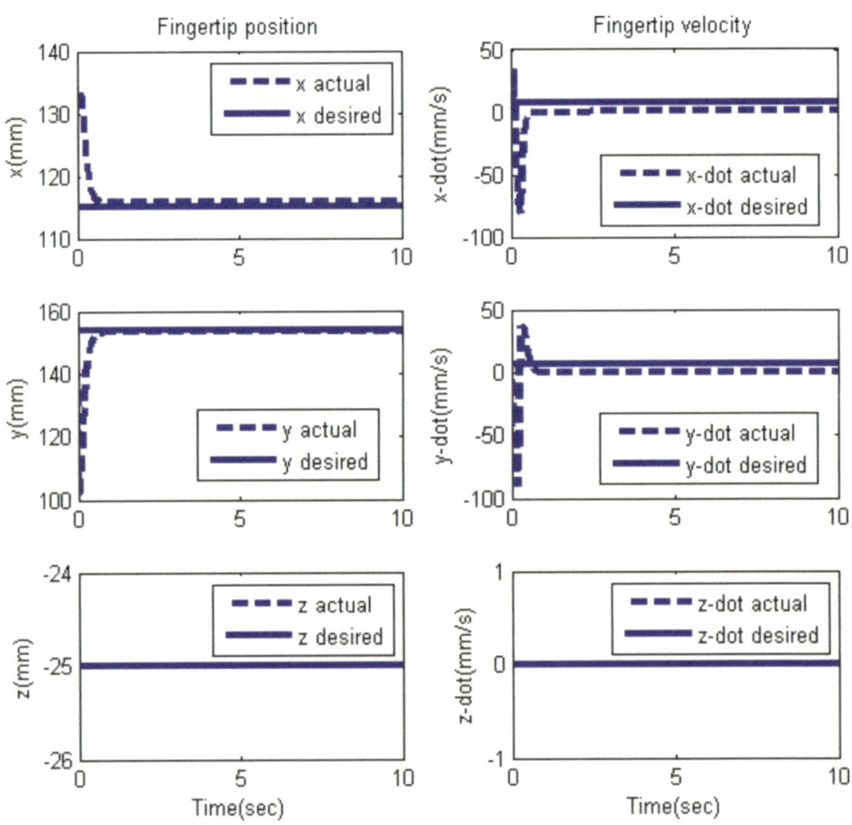

Figure 10.
Motion of fingertip 1 without impedance controller.

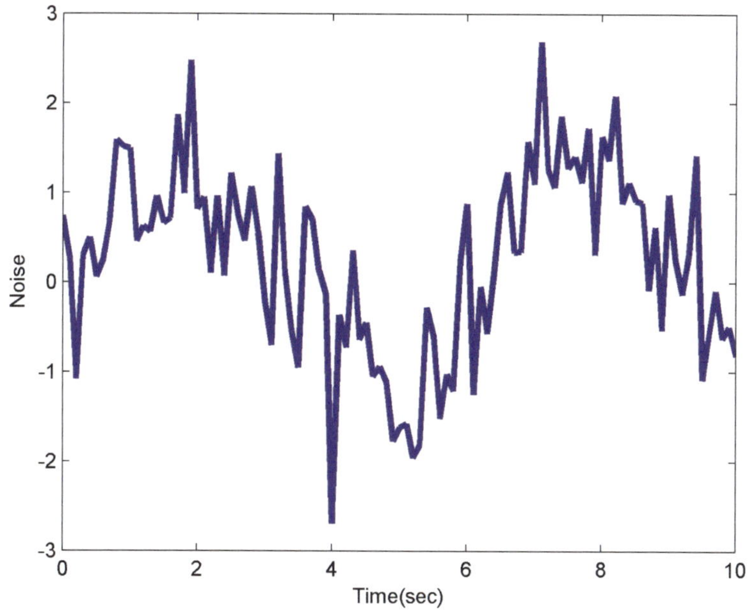

Figure 11.
Contact force signal with a white Gaussian noise.

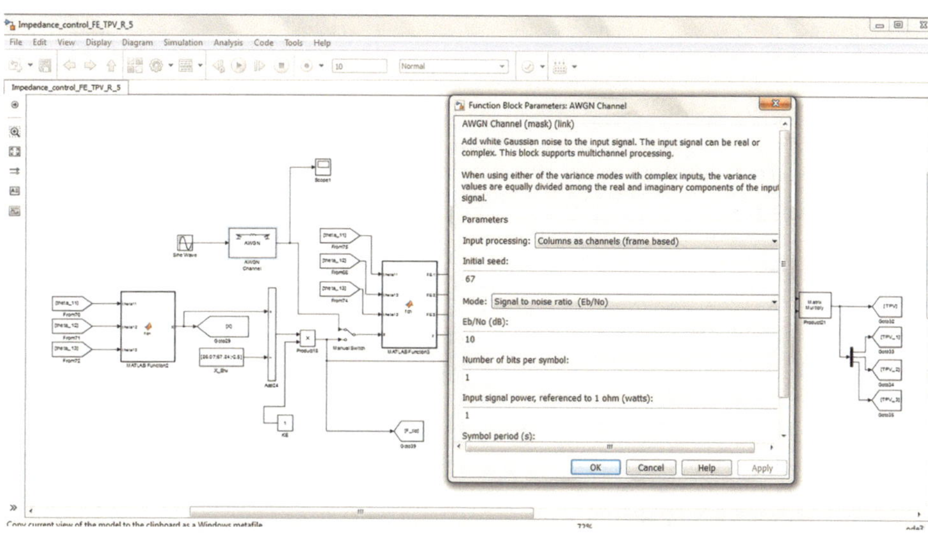

Figure 12.
Parameters of the applied white Gaussian noise.

in the test of Section 4.2. The disturbances on the robotic hand cause change of the fingertip contact forces. Here, we assumed that the disturbances result in a contact force with a white Gaussian noise [29–32] shown in **Figure 11**. The Gaussian noise model is generated using block AWGN channel in Simulink. The position of the AWGN channel is shown in **Figure 5**. The parameters of white Gaussian noise are set as shown in **Figure 12**. The designed controllers have achieved the functionality of rejecting the disturbances and kept the motion of the finger as shown in **Figure 13**.

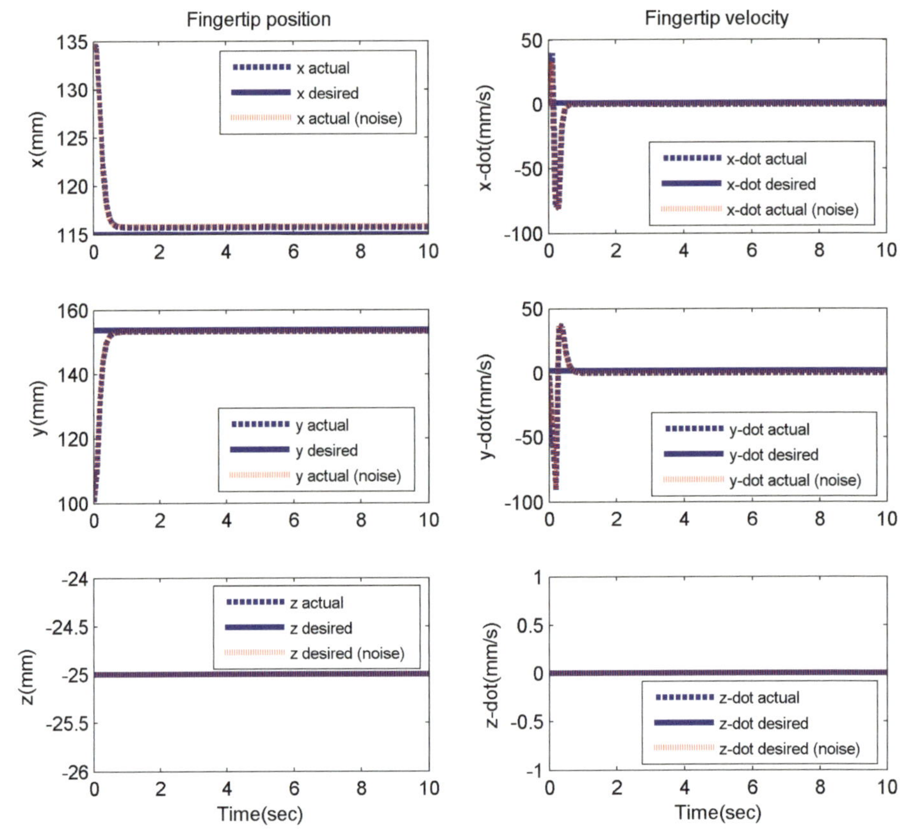

Figure 13.
Response of the designed controllers to the disturbances.

5. Conclusion

Grasping by two fingertips plays a distinctive role in its performance related to robotic hand-environment interaction. A mathematical model including the dynamic equation has been derived based on the Lagrange technique. The coupling between two fingers and the fingertip contacts with unknown grasped objects has been presented in this model. The performance of the derived model is confirmed when a designed based model controller is implemented in Simulink. A computed torque controller of six PD controllers has been implemented to control the rotation of joints. The impedance controller is applied to operate as the outer loop of the overall control system. This controller allowed the regulation of the fingertip contact forces without overshoot response, which is essential for safe grasping, especially in holding fragile objects. Besides this, the impedance gains have been obtained using a genetic algorithm with a new behavior of improving the response of the position and velocity errors. The applied genetic algorithm method has minimized the position error and the velocity error while keeping the operation of the overall control system. The robot finger has been ordered to follow a specific joint path and fingertip position grasping; the results of simulations have achieved a typical accomplishment of trajectory tracking. Finally, the overall controllers have shown a perfect rejection of disturbances in terms of fingertip contact force. As a future work, it is recommended to apply the proposed controllers of this study to mobile manipulator for grasping and manipulation objects.

Appendix A. Calculating the representation of Q_i,

Let

$$\vec{F_1} = -m_1 g\vec{j}, \vec{F_2} = -m_2 g\vec{j}, \vec{F_3} = -m_3 g\vec{j},$$

and

$$\vec{r}_1 = x_1\vec{i} + y_1\vec{j}, \vec{r}_2 = x_2\vec{i} + y_2\vec{j}, \vec{r}_3 = x_3\vec{i} + y_3\vec{j}.$$

Then

$$\vec{r}_1 = d_1\cos\theta_1\vec{i} + d_1\sin\theta_1\vec{j},$$

$$\vec{r_2} = (L_1\cos\theta_1 + d_2\cos(\theta_1+\theta_2))\vec{i} + (L_1\sin\theta_1 + d_2\sin(\theta_1+\theta_2))\vec{j},$$

$$\begin{aligned}\vec{r}_3 &= (L_1\cos\theta_1 + L_2\cos(\theta_1+\theta_2) + d_3\cos(\theta_1+\theta_2+\theta_3))\vec{i}\\ &+(L_1\sin\theta_1 + L_2\sin(\theta_1+\theta_2) + d_3\sin(\theta_1+\theta_2+\theta_3))\vec{j}.\end{aligned}$$

The partial derivatives of $\vec{r_i}$ with respect to θ_i are

$$\frac{\partial\vec{r}_1}{\partial\theta_1} = -d_1\sin\theta_1\vec{i} + d_1\cos\theta_1\vec{j},$$

$$\frac{\partial\vec{r}_1}{\partial\theta_2} = 0,$$

$$\frac{\partial\vec{r}_1}{\partial\theta_3} = 0,$$

$$\frac{\partial\vec{r}_2}{\partial\theta_1} = [-L_1\sin\theta_1 - d_2\sin(\theta_1+\theta_2)]\vec{i} + [L_1\cos\theta_1 + d_2\cos(\theta_1+\theta_2)]\vec{j},$$

$$\frac{\partial\vec{r}_2}{\partial\theta_2} = -d_2\sin(\theta_1+\theta_2)\vec{i} + d_2\cos(\theta_1+\theta_2)\vec{j},$$

$$\frac{\partial\vec{r}_2}{\partial\theta_3} = 0,$$

$$\begin{aligned}\frac{\partial\vec{r}_3}{\partial\theta_1} &= (-L_1\sin\theta_1 - L_2\sin(\theta_1+\theta_2) - d_3\sin(\theta_1+\theta_2+\theta_3))\vec{i}\\ &+ (L_1\cos\theta_1 + L_2\cos(\theta_1+\theta_2) + d_3\cos(\theta_1+\theta_2+\theta_3))\vec{j},\end{aligned}$$

$$\frac{\partial\vec{r}_3}{\partial\theta_2} = (-L_2\sin(\theta_1+\theta_2) - d_3\sin(\theta_1+\theta_2+\theta_3))\vec{i} + (L_2\cos(\theta_1+\theta_2) + d_3\cos(\theta_1+\theta_2+\theta_3))\vec{j},$$

$$\frac{\partial\vec{r}_3}{\partial\theta_3} = -d_3\sin(\theta_1+\theta_2+\theta_3)\vec{i} + d_3\cos(\theta_1+\theta_2+\theta_3))\vec{j},$$

The matrices' elements in Eq. (28) are

$$\begin{aligned}m_{11} &= m_1{d_1}^2 + I_1 + m_2{L_1}^2 + m_2{d_2}^2 + 2m_2L_1d_2\cos\theta_2 + I_2 + m_3{L_1}^2 + m_3{L_2}^2\\ &+ 2m_3L_1L_2\cos\theta_2 + 2m_3L_1d_3\cos(\theta_2+\theta_3) + 2m_3L_2d_3\cos\theta_3 + m_3{d_3}^2 + I_3,\end{aligned}$$

$$\begin{aligned}m_{12} &= m_2{d_2}^2 + m_2L_1d_2\cos\theta_2 + I_2 + m_3L_1L_2\cos\theta_2 + m_3L_1d_3\cos(\theta_2+\theta_3) + 2m_3{L_2}^2\\ &+ 2m_3L_2d_3\cos\theta_3 + m_3{d_3}^2 + I_3,\end{aligned}$$

$$m_{13} = m_3L_1d_3\cos(\theta_2+\theta_3) + m_3L_2d_3\cos\theta_3 + m_3{d_3}^2 + I_3,$$

$$m_{21} = m_2{d_2}^2 + m_2L_1d_2\cos\theta_2 + I_2 + m_3L_1L_2\cos\theta_2 + m_3L_1d_3\cos(\theta_2+\theta_3)$$
$$+2m_3L_2d_3\cos\theta_3 + 2m_3{L_2}^2 + m_3{d_3}^2 + I_3,$$

$$m_{22} = m_2{d_2}^2 + I_2 + m_3{L_2}^2 + 2m_3L_2d_3\cos\theta_3 + m_3{d_3}^2 + I_3,$$

$$m_{23} = m_3L_2d_3\cos\theta_3 + m_3{d_3}^2 + I_3,$$

$$m_{31} = m_3L_1d_3\cos(\theta_2+\theta_3) + m_3L_2d_3\cos\theta_3 + m_3{d_3}^2 + I_3,$$

$$m_{32} = m_3L_2d_3\cos\theta_3 + m_3{d_3}^2 + I_3,$$

$$m_{33} = m_3{d_3}^2 + I_3,$$

$$d_{11} = -2m_2L_1d_2\sin\theta_2\dot{\theta}_2 - 2m_3L_1L_2\sin\theta_2\dot{\theta}_2 - 2m_3L_1d_3\sin(\theta_2+\theta_3)\dot{\theta}_2$$
$$-2m_3L_1d_3\sin(\theta_2+\theta_3)\dot{\theta}_3 - 2m_3L_2d_3\sin\theta_3\dot{\theta}_3,$$

$$d_{12} = -m_2L_1d_2\sin\theta_2\dot{\theta}_2 - m_3L_1L_2\sin\theta_2\dot{\theta}_2 - m_3L_1d_3\sin(\theta_2+\theta_3)\dot{\theta}_2$$
$$-m_3L_1d_3\sin(\theta_2+\theta_3)\dot{\theta}_3 - 2m_3L_2d_3\sin\theta_3\dot{\theta}_3 - m_3L_1d_3\sin(\theta_2+\theta_3)\dot{\theta}_3,$$

$$d_{13} = -m_3L_1d_3\sin(\theta_2+\theta_3)\dot{\theta}_3 - m_3L_2d_3\sin\theta_3\dot{\theta}_3,$$

$$d_{21} = -m_2L_1d_2\sin\theta_2\dot{\theta}_2 - m_3L_1L_2\sin\theta_2\dot{\theta}_2 - m_3L_1d_3\sin(\theta_2+\theta_3)\left[\dot{\theta}_2+\dot{\theta}_3\right]$$
$$-2m_3L_2d_3\sin\theta_3\dot{\theta}_3 + m_2L_1d_2\sin\theta_2\dot{\theta}_1 + m_2L_1d_2\sin\theta_2\dot{\theta}_2 + m_3L_1L_2\sin\theta_2\dot{\theta}_1$$
$$+m_3L_1d_3\sin(\theta_2+\theta_3)\dot{\theta}_1 + m_3L_1L_2\sin\theta_2\dot{\theta}_2 + m_3L_1d_3\sin(\theta_2+\theta_3)\dot{\theta}_2 + m_3L_1d_3\sin(\theta_2+\theta_3)\dot{\theta}_3,$$

$$d_{22} = -2m_3L_2d_3\sin\theta_3\dot{\theta}_3,$$

$$d_{23} = -m_3L_2d_3\sin\theta_3\dot{\theta}_3,$$

$$d_{31} = -m_3L_1d_3\sin(\theta_2+\theta_3)\left[\dot{\theta}_2+\dot{\theta}_3\right] - m_3L_2d_3\sin\theta_3\dot{\theta}_3 + m_3L_1d_3\sin(\theta_2+\theta_3)\dot{\theta}_1$$
$$+m_3L_2d_3\sin\theta_3\dot{\theta}_1 + m_3L_1d_3\sin(\theta_2+\theta_3)\dot{\theta}_2 + 2m_3L_2d_3\sin\theta_3\dot{\theta}_2$$
$$+m_3L_1d_3\sin(\theta_2+\theta_3)\dot{\theta}_3 + m_3L_2d_3\sin\theta_3\dot{\theta}_3,$$

$$d_{32} = -m_3L_2d_3\sin\theta_3\dot{\theta}_3 + m_3L_2d_3\sin\theta_3\dot{\theta}_2 + m_3L_2d_3\sin\theta_3\dot{\theta}_3,$$

$$d_{33} = 0,$$

$$g_1 = m_1gd_1\cos\theta_1 + m_2gL_1\cos\theta_1 + m_2gd_2\cos(\theta_1+\theta_2) + m_3gL_1\cos\theta_1$$
$$+m_3gL_2\cos(\theta_1+\theta_2) + m_3gd_3\cos(\theta_1+\theta_2+\theta_3),$$

$$g_2 = m_2gd_2\cos(\theta_1+\theta_2) + m_3gL_2\cos(\theta_1+\theta_2) + m_3gd_3\cos(\theta_1+\theta_2+\theta_3),$$

$$g_3 = m_3gd_3\cos(\theta_1+\theta_2+\theta_3),$$

Author details

Izzat Al-Darraji[1,2]*, Ali Kılıç[1] and Sadettin Kapucu[1]

1 Mechanical Engineering Department, University of Gaziantep, Gaziantep, Turkey

2 Automated Manufacturing Department, University of Baghdad, Baghdad, Iraq

*Address all correspondence to: dr.izzat79@gmail.com

References

[1] Goodrich MA, Schultz AC. Human–robot interaction: A survey. Foundations and Trends in Human–Computer Interaction. 2008;**1**:203-275

[2] Green SA et al. Human-robot collaboration: A literature review and augmented reality approach in design. International Journal of Advanced Robotic Systems. 2008;**5**:1-18

[3] Ramaswamy CVV, Deborah AS. A survey of robotic hand-arm systems. International Journal of Computer Applications. 2015;**109**:26-31

[4] Cutkosky MR. Robotic grasping and fine manipulation. Boston, MA: Springer; 1985

[5] Arimoto S et al. Modeling and control for 2-D grasping of an object with arbitrary shape under rolling contact. SICE Journal of Control, Measurement, and System Integration. 2009;**2**:379-386

[6] Corrales JA. et al. Modeling and simulation of a multi-fingered robotic hand for grasping tasks. In: 11th International Conference on Control Automation Robotics & Vision. Singapore; 2010. pp. 1577-1582

[7] Boughdiri R, et al. Dynamic modeling of a multi-fingered robot hand in free motion. In: Eighth International Multi-Conference on Systems, Signals & Devices. Sousse; 2011. pp. 1-7

[8] Boughdiri R et al. Dynamic modeling and control of a multi-fingered robot hand for grasping task. Procedia Eng ineering. 2012;**41**:923-931

[9] Lewis FL et al. Robot manipulator control: Theory and practice. New York, USA: CRC Press Publisher. 2003

[10] Lu WS, Meng QH. Impedance control with adaptation for robotic manipulators. IEEE Transactions on Robotics and Automation. 1991;7: 408-415

[11] Hogan N. Impedance control: An approach to manipulation: Part I-III. ASME Journal of Dynamic Systems, Measurement, and Control. 1985;**107**: 1-24

[12] Hogan N. Stable execution of contact tasks using impedance control. In: Proceedings of IEEE International Conference on Robotics and Automation, Raleigh, NC, USA; 1987. pp. 1047-1054. DOI:10.1109/ROBOT.1987.1087854

[13] Chen Z, et al. Experimental study on impedance control for the five-finger dexterous robot hand DLR-HIT II. In: 2010 IEEE/RSJ International Conference on Intelligent Robots and Systems; Taipei, 2010. pp. 5867-5874. DOI: 10.1109/IROS.2010.5649356

[14] Huang J et al. Method of grasping control by computing internal and external impedances for two robot fingers, and its application to admittance control of a robot hand-arm system. International Journal of Advanced Robotic Systems. 2014;**12**: 1-11

[15] Zhang T et al. Development and experimental evaluation of multi-fingered robot hand with adaptive impedance control for unknown environment grasping. Robotica. 2016; **34**:1168-1185

[16] Murray RM et al. Mathematical Introduction to Robotic Manipulation. USA: CRC Press; 1994

[17] Bruyninckx H. Robot Kinematics and Dynamics. Leuven, Belgium: Katholieke Universiteit Leuven, Department of Mechanical Engineering; 2010

[18] Fabien B. Analytical System Dynamics Modeling and Simulation. New York, USA: Springer US; 2009

[19] Rubio JJ. Robust feedback linearization for nonlinear processes control. ISA Transactions. 2018. DOI: 10.1016/j.isatra.2018.01.017

[20] Aguilar-Ibañez C. Stabilization of the PVTOL aircraft based on a sliding mode and a saturation function. International Journal of Robust and Nonlinear Control. 2017;**27**:4541-4553

[21] Rubio JJ et al. Control of two electrical plants. Asian Journal of Control. 2017;**20**:1504-1518. DOI: 10.1002/asjc.1640

[22] Aguilar-Ibañez C, Sira-Ramirez, Hebertt, Acosta, JÁ. Stability of active disturbance rejection control for uncertain systems: A Lyapunov perspective: An ADRC Stability Analysis.. International Journal of Robust and Nonlinear Control. 2017;**27**: 4541-4553. DOI: 10.1002/rnc.3812

[23] Llama MA et al. Stable computed-torque control of robot manipulators via fuzzy self-tuning. IEEE Transactions on Systems, Man, and Cybernetics Part B (Cybernetics). 2000;**30**:143-150

[24] Santos WMD, Siqueira AAG. Impedance control of a rotary series elastic actuator for knee rehabilitation. IFAC Proceedings Volumes. 2014;**47**: 4801-4806

[25] Zhao Y et al. Impedance control and performance measure of series elastic actuators. IEEE Transactions on Industrial Electronics. 2018;**65**: 2817-2827

[26] Mehdi H, Boubaker O. Stiffness and impedance control using Lyapunov theory for robot-aided rehabilitation. International Journal of Social Robotics. 2012;**4**:107-119

[27] Gen M, Cheng R. Genetic Algorithms and Engineering Design. New York, USA: Wiley; 1997

[28] Sivanandam SN, Deepa SN. Introduction to Genetic Algorithms. Berlin: Springer; 2008

[29] Azizi A. Computer-based analysis of the stochastic stability of mechanical structures driven by white and colored noise. Sustainability (Switzerland). 2018;**10**(10):3419

[30] Azizi A, Yazdi PG. White noise: Applications and mathematical modeling. In: Computer-Based Analysis of the Stochastic Stability of Mechanical Structures Driven by White and Colored Noise. Singapore: Springer; 2019. pp. 25-36

[31] Azizi A, Yazdi PG. Modeling and control of the effect of the noise on the mechanical structures. In: Computer-Based Analysis of the Stochastic Stability of Mechanical Structures Driven by White and Colored Noise. Singapore: Springer; 2019. pp. 75-93

[32] Azizi A, Yazdi PG. Introduction to noise and its applications. In: Computer-Based Analysis of the Stochastic Stability of Mechanical Structures Driven by White and Colored Noise

Chapter 11

Research on Key Quality Characteristics of Electromechanical Product Based on Meta-Action Unit

Yan Ran, Xinlong Li, Shengyong Zhang and Genbao Zhang

Abstract

Electromechanical products have many quality characteristics, representing their quality. In addition, there are long-existed quality problems of electromechanical products, such as poor accuracy, short precision life, large fluctuations in performance, frequently failing, and so on. Based on meta-action unit (MU) for electromechanical products, this book chapter proposes a key quality characteristic control method, which provides theoretical and technical support for essentially guaranteeing the complete machine's quality. The formation mechanisms of MU's four key quality characteristics (precision, precision life, performance stability, and reliability) are studied. Moreover, we introduce an overview of key quality characteristic control methods based on MU. The complex large system research method of "decomposition-analysis-synthesis" is adopted to study these key science problems.

Keywords: key quality characteristics, electromechanical product, meta-action unit

1. Introduction

The construction of industrial modernization cannot be separated from the electromechanical integration [1]. Electromechanical products play an extremely important role in the current society. The technical level of electromechanical products determines the development level of the national economy and also directly reflects the country's manufacturing capacity, scientific and technological strength, economic strength, international competitiveness, and other comprehensive national strengths [2]. Poor quality of electromechanical products will not only bring economic losses to enterprises but also seriously affect the international market competitiveness of products and, more seriously, the overall image and economic strength of the country. Therefore, accelerating the development of electromechanical products and improving the quality of products are of vital significance for the national economic construction, national defense security, and social stability.

There are many quality indexes of electromechanical products. Among them, the four indexes (precision, precision life, performance stability, and reliability) are the most concerned by users, which together constitute the key quality characteristics of electromechanical products and decisively affect the market competitiveness of electromechanical products. Therefore, in order to occupy a place in the competitive high-end market, we must first improve the level of these key quality characteristics [3]. Therefore, carrying out a research into the electromechanical product quality control technology, especially, the quality of its key features, is extremely important. However, it is very difficult to control the quality of complex electromechanical products, not only for its complex structure but also for its working process, a dynamic process composed of many motions. The commonly used structural decomposition methods are mostly the static research based on the quality characteristics of parts in the design process. At the same time, there is a lot of uncertainty and coupling in the process of motion, which makes the dynamic quality characteristics greatly different from the static ones. In view of these problems, this paper puts forward the research idea of quality characteristic control technology of electromechanical product based on MU [4].

2. Literature review

There are many quality characteristics of electromechanical products, but its key characteristics mainly include precision, precision life, performance stability, and reliability, which comprehensively reflect the availability of the product. At present, the analysis of precision is quite common for the stages of design, manufacture, and operation of electromechanical products, but the research of precision life, performance stability, and reliability is few.

The precision of electromechanical products includes geometric accuracy, motion accuracy, transmission accuracy, position accuracy, etc., among which geometric accuracy is the basis and guarantee of other precisions. There are many studies on the precision of electromechanical products. Enterprises pay more attention to the precision of electromechanical products, especially for machine tools. Sata et al. put forward a method to improve machining accuracy of machining center through computer control compensation earlier [5]. Ceglarek et al. used the state space model to model the stream of variation (SOV) of the multi-stage manufacturing system, described the errors of process parameters and product quality characteristics with the vector tolerance, and applied it in automobile body-in-white assembly and automobile engine cylinder head machining [6]. Zeyuan et al. carried out simulation and experimental research on geometric error detection, identification, and compensation of translational axis of NC machine tools by two geometric error-modeling methods [7]. Li et al. summarized the characteristics of the existing error compensation methods and discussed the future development direction of improving the spatial positioning accuracy of five-axis CNC machine tools [8]. Yongwei et al. proposed a method for predicting the motion accuracy of CNC machine tools based on the time-series deep learning network [9].

The research on precision life mainly focuses on the working stage of electromechanical products, and the research object mainly focuses on the precision life of servo feed system (such as screw pair, etc.). Min et al. carried out a simulation test of the precision retaining ability degradation of the ball screw pair to achieve a fast prediction of the life of the ball screw pair [10]. Liping et al. conducted a gray prediction of the rotary precision life of the spindle of CNC lathe [11]. Linlin studied the precision retaining ability of servo feed system of CNC machine tool and its test method [12]. Hu et al. proposed a vibration aging process parameter selection method based on modal analysis and harmonious response analysis to improve the precision life of large basic parts of CNC machine tools in view of the nonstandard phenomenon of vibration aging process in domestic manufacturing enterprises [13].

Shijun et al. analyzed the influence of axial load on friction torque of ball screw pair through experiments and pointed out that the pre-tightening force plays a leading role in the fluctuation of friction torque of ball screw pair [14].

The research on performance stability mainly focuses on the research of performance index and parameter design. Saitou et al. summarized the development history of structural design and optimization of mechanical products and optimized the parameters to achieve the purpose of improving the stability of its performance [15]. Ta et al. studied the influence of friction on sliding stability of multibody systems [16]. Caro et al. improved the robust design method based on the sensitive area and proposed the comprehensive robust design method of mechanism deviation [17]. The size of the mechanism was calculated according to the performance robustness index, and the optimal deviation of the mechanism was calculated using the comprehensive deviation method. Gao et al. analyzed the robustness and robustness estimation of the product quality features [18]. They established a mathematical model based on minimum sensitivity region estimation (MSRE) for the robust optimization of the product quality features. Hui et al. constructed a dual-inertia feed system model based on the sensitivity of the closed-loop transfer function and analyzed the influence of the change of feed system stiffness on the stability of motion accuracy [19].

The research on reliability mainly focuses on reliability design, assembly reliability control, and fault diagnosis technology. Based on the shortcomings of traditional stress-strength interference model, Zhang et al. modeled the reliability from a dynamic perspective [20]. Pinghua et al. studied the assembly reliability control technology of electromechanical products based on fault modeling of motion unit to improve the reliability of assembly process [21]. Genbao et al. proposed the assembly process and technology driven by reliability and used the dynamic Bayesian method to model and control the assembly process [22]. Assaf and Dugan proposed a diagnostic method for large systems using monitors and sensors for reliability analysis by optimizing the diagnostic decision tree (DDTs) [23]. Yu et al. proposed a fault maintenance strategy for field equipment of assembly system [24]. And according to the differences in reliability design of different types of equipment, a comprehensive maintenance mode was introduced to ensure the reliability of automatic docking assembly of large aircraft parts. Qinghu et al. carried out a systematic analysis and prediction modeling research on the main failure mechanism and failure evolution rule of key components of the power transmission system [25].

Quality characteristic control is a comprehensive technical control activity based on decoupling or prediction of quality characteristic, including analysis of quality characteristic at each granularity and adjustment of influencing factors. Moreover, according to quality fluctuation, quality characteristic is under control. Haifeng studied the coupling mapping relation of quality characteristics of complex electromechanical products [26]. Then, combined with axiomatic design principle, they proposed the idea of decoupling design of quality characteristics of complex electromechanical products and presented the decoupling control model in the process of quality characteristic mapping of complex electromechanical products. Xianghua proposed a multiscale collaborative and intelligent quality control method for key parts of air separation equipment including the air compressor, the turbine expander, and the whole machine and conducted an in-depth research on coupling mapping and collaborative control of quality characteristics of large-scale air separation equipment from the whole machine to each part at all scales [27]. Nada studied the quality prediction in the design process of manufacturing system, constructed a basic framework on the basis of the manufacturing system configuration parameters to evaluate the quality of the products, and constructed a comprehensive evaluation model of system configuration based on quality by analytic hierarchy process (AHP), which can transform the structural parameters of manufacturing system into configuration capability indexes and then predict and control them [28]. Xianlin conducted an in-depth research on the quality characteristic prevention and control key technologies such as the quality characteristic evolution mechanism, coupling and decoupling control, timing data prediction, immune diagnosis, and control in the manufacturing process of electromechanical products [29]. In order to solve the coupling problem in quality characteristic predictive control, Zhentao established a quality characteristic predictive decoupling model based on predictive control theory [30]. Yang et al. established and solved the predictive control model based on particle swarm optimization algorithm and support vector machine on the basis of analyzing the characteristics of multiprocess and multistage product quality prediction control and realized the global optimization of multistage product quality prediction and related process parameters [31].

3. Findings

Through the analysis of the above research status, it can be seen that the following main problems exist in the quality characteristic control of electromechanical products at present.

Traditional quality control is mainly based on the quality control of parts, precisely, the precision, However, the combination with the function of parts is far from the expectance. Particularly, general electromechanical products exist for the realization of a certain motion function, and the overall functional failure of the product is caused by the inability to realize the basic motion function. Therefore, quality control should be placed in the basic level of motion, as long as the basic motion function is normal and the overall function of the product is guaranteed. Starting from the quality control of MUs, the quality control of the whole machine is realized through the quality control of the unit level. This control mode is more simple and effective, which does not need to simplify the model and is conducive to the accomplishment of refined quality control.

In terms of the key quality characteristic control technology, there are many researches on the pure precision but few achievements on the precision life, performance stability, reliability, and their coupling control. In fact, the quality gap of electromechanical products at present mainly lies in the aspects of precision life, performance stability, and reliability, among which reliability is the biggest bottleneck of electromechanical products. How to improve the precision life, performance stability, and reliability of electromechanical products as soon as possible is an urgent problem to be solved in the manufacturing industry.

4. Formation mechanisms of MU's four key quality characteristics

Meta-action unit (MU) is a kind of basic action unit of electromechanical products decomposed in accordance with failure model analysis (FMA) structural decomposition method. It has the specific function of independently completing the specified movement or operation. Compared with other decomposition units, MU is more suitable for quality and reliability analysis. However, the complex quality characteristics are reflected not only in the MU's motion accuracy but also in its precision life, performance stability, reliability, and other aspects. So it is necessary to conduct an in-depth study on the formation mechanism of each quality characteristic.

The quality characteristics of electromechanical products are very numerous, generally including 12 aspects, i.e., functional compliance, performance stability, application reliability, precision retaining ability, technology primacy, cost economy, esthetic appearance, maintenance convenience, and service timeliness, character uniqueness, security assurance, perception satisfaction, etc., which lead to the resultant diversity of MU's quality characteristics. During the research process, the key quality characteristics need to be extracted from the perspective of the user's requirements and the fault distribution. At the same time, as the smallest motion unit decomposed from electromechanical products, most of MUs' quality characteristics are dynamic; therefore, it is necessary to conduct an in-depth study on their formation mechanism and extract the characteristic indicators of each key quality characteristic from the perspective of motion. Moreover, to expand the amount of the sample data, the corresponding models are established and applied to the similarity determination of MUs.

4.1 Formation mechanism of the MU motion accuracy

Accuracy is the degree to which actual geometric parameters conform to ideal ones, and their deviation values are often defined as errors. The accuracy of electromechanical products generally includes geometric accuracy, motion accuracy, transmission accuracy, and positioning accuracy [32]. For MUs, the accuracy refers to the motion accuracy of the output part that implements the meta-action. Moreover, the movement accuracy of the unit actuator is also affected by the manufacture and assembly errors of other parts in the unit (support, transmission, and fasteners). For example, as the actuator of the rack moves the MU, the translation accuracy of the rack is the only measure indicator. In the actual situation, there are many kinds of error sources that affect the motion accuracy of the MU actuators. Actually, the formation process of the accuracy is also the transmission process of the errors.

4.1.1 The MU motion accuracy

For translation and rotation MU, the motion accuracy of the actuator also includes translation and rotation accuracy. The corresponding translation accuracy indicators of the actuator mainly include positioning accuracy, repeat positioning accuracy, reverse error, moving straightness (trajectory), moving parallelism (trajectory), and resolution (minimum translation increment). The rotation accuracy indicators include positioning accuracy, repeat positioning accuracy, indexing accuracy, reverse error, rotation face fluctuation (radial and end faces), and rotation fluctuation (axial).

The motion accuracy of the unit actuator is also affected by the geometric and assembly accuracy of the various parts in the unit. The geometric accuracy of the part includes the geometric position and shape accuracy, expressed as static errors such as size, position, and shape. According to ISO230-1:1996 and GB/T1800.1-2009 "Limit and fit Part 1: The basis of tolerance, deviation and fit," the size and position error of the MU actuator refers to the defects in the design, manufacture, and assembly process, which cause the MU actuator (output member) deviating from the ideal size and position. It mainly includes distance dimension error, diameter dimension error, angular dimension error, parallelism, coaxiality, verticality, position, symmetry, and round and full runout. The geometric shape error of the MU actuator refers to the deviation between the actual and ideal shape after machining, which is set as a machining range of the normal distribution and mainly includes straightness, flatness, roundness, cylindricity, line profile, and surface profile.

4.1.2 Error model of the MU motion accuracy

4.1.2.1 Topology structure description of electromechanical product

The electromechanical product is a kind of multibody system, a complete abstract and effective description of general complicated mechanical system, which is the optimal mode of analyzing and researching complicated mechanical system [33–35]. The multibody kinematic theory is used to establish the synthetic space motion error model of the MU actuator in this chapter.

An MU is regarded as a body, and there is only a single degree-of-freedom relative motion with the constraint type of translation or rotation. The motion of electromechanical products, open or closed loop, aims to accomplish a corresponding function. Because closed-loop motion can be converted as open with specific constraint, the topological structure for general electromechanical products is expressed as follows (**Figure 1**). The inertial reference system and MU are regarded as different bodies. Then the bodies are numbered in turn along the direction away, according to the natural growth sequence, from one branch to another (each branch accomplishing its corresponding function).

The low-order body array is used to describe the topological structure of the multibody system [36], which can be obtained by the following calculation formula (1). The optional body B_k is any typical body in the system, and the sequence number of its n-th lower-order body is defined as

$$L^n(k) = j \tag{1}$$

where L is the low-order body operator and body B_k is the n-th high-order body of body B_j (**Table 1**). $L^n(k) = L(L^{n-1}(k))$, $L^0(k) = k$, and $L^n(k) = 0$. Moreover, when body B_j is the adjacent lower-order body of body B_k, $L(k) = j$.

4.1.2.2 Motion error characteristic matrix of adjacent MUs

According to the MU motion of the electromechanical product in the actual working process, the description method of multibody system error is used to study

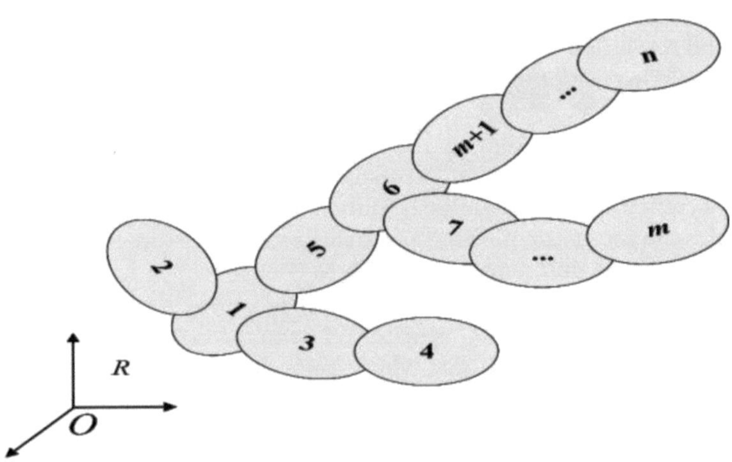

Figure 1.
The topological structure diagram of electromechanical products based on MUs.

k	1	2	3	4	5	6	7	…	m	$m+1$	…	n
$L^0(k)$	1	2	3	4	5	6	7	…	m	$m+1$	…	n
$L^1(k)$	0	1	1	3	1	5	6	…	$m-1$	6	…	$n-1$
$L^2(k)$	0	0	0	1	0	1	5	…	$m-2$	5	…	$n-2$
$L^3(k)$	0	0	0	0	0	0	1	…	$m-3$	1	…	$n-3$
$L^m(k)$	0	0	0	0	0	0	0	0	0	0	0	0
$L^n(k)$	0	0	0	0	0	0	0	0	0	0	0	0

Table 1.
The low-order bodies of electromechanical products based on MU.

the motion error of the MU actuator (**Figure 2**). P_k, P_k^I, and P_k^E are the actual position vector, the ideal position vector, and the position error vector of the origin Q_k of the body B_k motion reference coordinate system in its low-order body B_j, respectively; and D_k, D_k^I, and D_k^E are the actual displacement vector, the ideal displacement vector, and the displacement error vector of the origin O_k of body B_k reference coordinate system.

From the kinematic point of view, the errors of MU actuators are divided into two categories. One is the errors that happened in the motion reference coordinate of MU (body B_k), described with the position (static) error vector P_k^E and the position error feature (transformation) matrix T_{jk}^{PE}. Another is the motion process error (translation or rotation) of the MU (body B_k) relative to the actual motion reference coordinate, described with the translation (motion) error vector D_k^E (linear or angular displacement) and the motion error feature (transformation) matrix T_{jk}^{DE}.

4.1.2.2.1 Motion error characteristic matrix from translation MU to translation MU

When the MU (body B_k) translates x_{jk}^D, y_{jk}^D, and z_{jk}^D along the X-axis, Y-axis, and Z-axis relative to its adjacent MU (lower-order body B_j), its actuator's translation

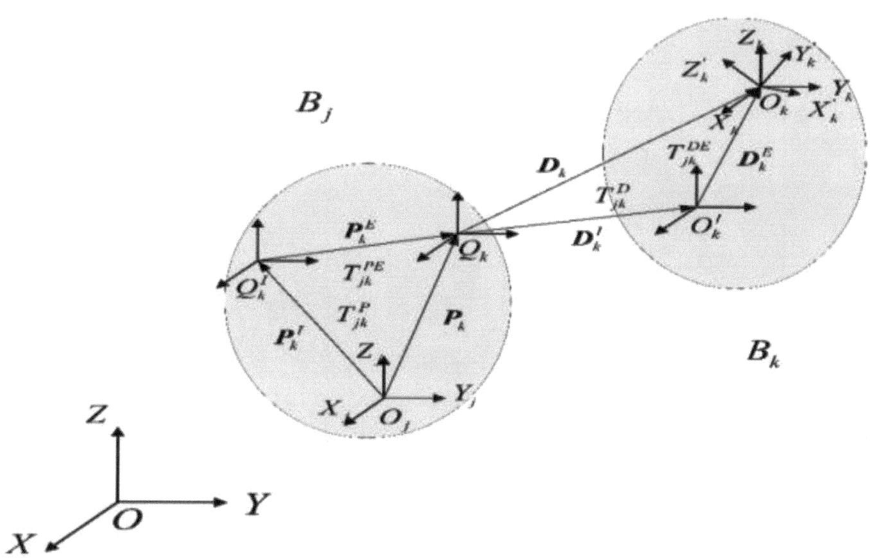

Figure 2.
The relative motion sketch of adjacent MUs.

matrices and translation error feature matrices can be expressed, respectively, as follows:

$$T^{D}_{\rm jk}(x)=\begin{bmatrix}1&0&0&x^{D}_{\rm jk}\\0&1&0&0\\0&0&1&0\\0&0&0&1\end{bmatrix},T^{\rm DE}_{\rm jk}(\Delta{\rm x})=\begin{bmatrix}1&0&0&\Delta x^{\rm DE}_{\rm jk}\\0&1&0&0\\0&0&1&0\\0&0&0&1\end{bmatrix}$$

$$T^{D}_{\rm jk}(y)=\begin{bmatrix}1&0&0&0\\0&1&0&y^{D}_{\rm jk}\\0&0&1&0\\0&0&0&1\end{bmatrix},T^{\rm DE}_{\rm jk}(\Delta{\rm y})=\begin{bmatrix}1&0&0&0\\0&1&0&\Delta y^{\rm DE}_{\rm jk}\\0&0&1&0\\0&0&0&1\end{bmatrix}$$

$$T^{D}_{\rm jk}(z)=\begin{bmatrix}1&0&0&0\\0&1&0&0\\0&0&1&z^{D}_{\rm jk}\\0&0&0&1\end{bmatrix},T^{\rm DE}_{\rm jk}(\Delta{\rm z})=\begin{bmatrix}1&0&0&0\\0&1&0&0\\0&0&1&\Delta z^{\rm DE}_{\rm jk}\\0&0&0&1\end{bmatrix}$$

4.1.2.2.2 Motion error characteristic matrix from rotation MU to rotation MU

When the MU (body B_k) rotates α^{D}_{jk}, β^{D}_{jk}, and γ^{D}_{jk} (the Euler angle or the Karl single angle of the coordinate system k relative to the coordinate system j) along the X-axis, Y-axis, and Z-axis relative to its adjacent MU (lower-order body B_j), its actuator's rotation matrices and rotation error feature matrices can be expressed, respectively, as follows:

$$T^{D}_{\rm jk}(\alpha)=\begin{bmatrix}1&0&0&0\\0&cos\ \alpha^{D}_{\rm jk}&-sin\ \alpha^{D}_{\rm jk}&0\\0&sin\ \alpha^{D}_{\rm jk}&cos\ \alpha^{D}_{\rm jk}&0\\0&0&0&1\end{bmatrix},T^{\rm DE}_{\rm jk}(\Delta\alpha)=\begin{bmatrix}1&0&0&0\\0&cos\ \Delta\alpha^{\rm DE}_{\rm jk}&-sin\ \Delta\alpha^{\rm DE}_{\rm jk}&0\\0&sin\ \Delta\alpha^{\rm DE}_{\rm jk}&cos\ \Delta\alpha^{\rm DE}_{\rm jk}&0\\0&0&0&1\end{bmatrix}$$

$$T^{D}_{\rm jk}(\beta)=\begin{bmatrix}cos\ \beta^{D}_{\rm jk}&0&sin\ \beta^{D}_{\rm jk}&0\\0&1&0&0\\-sin\ \beta^{D}_{\rm jk}&0&cos\ \beta^{D}_{\rm jk}&0\\0&0&0&1\end{bmatrix},T^{\rm DE}_{\rm jk}(\Delta\beta)=\begin{bmatrix}cos\ \Delta\beta^{D}_{\rm jk}&0&sin\ \Delta\beta^{D}_{\rm jk}&0\\0&1&0&0\\-sin\ \Delta\beta^{D}_{\rm jk}&0&cos\ \Delta\beta^{D}_{\rm jk}&0\\0&0&0&1\end{bmatrix}$$

$$T^{D}_{\rm jk}(\gamma)=\begin{bmatrix}cos\ \gamma^{D}_{\rm jk}&-sin\ \gamma^{D}_{\rm jk}&0&0\\sin\ \gamma^{D}_{\rm jk}&cos\ \gamma^{D}_{\rm jk}&0&0\\0&0&1&0\\0&0&0&1\end{bmatrix},T^{\rm DE}_{\rm jk}(\Delta\gamma)=\begin{bmatrix}cos\ \Delta\gamma^{D}_{\rm jk}&-sin\ \Delta\gamma^{D}_{\rm jk}&0&0\\sin\ \Delta\gamma^{D}_{\rm jk}&cos\ \Delta\gamma^{D}_{\rm jk}&0&0\\0&0&1&0\\0&0&0&1\end{bmatrix}$$

Motion error characteristic matrix of synthetic motion at the MU (body B_k) translates first and then rotates relative to its adjacent MU (low-order body B_j). The specific motion order is translating x^{D}_{jk}, y^{D}_{jk}, and z^{D}_{jk} along the X-axis, Y-axis, and

Z-axis and rotating α_{jk}^{D} along the X-axis, β_{jk}^{D} along the Y-axis, and γ_{jk}^{D} along the Z-axis, and the ideal motion feature matrix of the unit actuator can be expressed as

$$T_{jk}^{D} = T_{jk}^{D}(x)T_{jk}^{D}(y)T_{jk}^{D}(z)T_{jk}^{D}(\alpha)T_{jk}^{D}(\beta)T_{jk}^{D}(\gamma) \tag{2}$$

The motion error feature matrix of the unit actuator is expressed as

$$T_{jk}^{DE} = T_{jk}^{DE}(\Delta x)T_{jk}^{DE}(\Delta y)T_{jk}^{DE}(\Delta z)T_{jk}^{DE}(\Delta \alpha)T_{jk}^{DE}(\Delta \beta)T_{jk}^{DE}(\Delta \gamma) \tag{3}$$

The position and position error feature matrix are expressed as

$$T_{jk}^{P} = T_{jk}^{P}(x)T_{jk}^{P}(y)T_{jk}^{P}(z)T_{jk}^{P}(\alpha)T_{jk}^{P}(\beta)T_{jk}^{P}(\gamma) \tag{4}$$

$$T_{jk}^{PE} = T_{jk}^{PE}(\Delta x)T_{jk}^{PE}(\Delta y)T_{jk}^{PE}(\Delta z)T_{jk}^{PE}(\Delta \alpha)T_{jk}^{PE}(\Delta \beta)T_{jk}^{PE}(\Delta \gamma) \tag{5}$$

4.1.2.3 Actual characteristic matrix of MUs

According to the geometric description method above, the actual feature (transformation) matrix of the MU actuator in the multibody system is obtained as

$$T_{jk} = T_{jk}^{P}T_{jk}^{PE}T_{jk}^{D}T_{jk}^{DE} \tag{6}$$

where T_{jk}, T_{jk}^{P}, T_{jk}^{PE}, T_{jk}^{D}, and T_{jk}^{DE} are the actual position, the ideal position, the static error, the ideal motion, and the motion error feature matrix of the MU actuator, respectively.

4.1.2.4 Integrated space motion errors of MUs

Set the homogeneous coordinates of the MU (body B_j) on any point W in subcoordinate system $O_j - X_jY_jZ_j$ as

$$W_j = \begin{bmatrix} w_{xj} & w_{yj} & w_{zj} & 1 \end{bmatrix}^T \tag{7}$$

Thus, the ideal homogeneous coordinates of the point W in subcoordinate system $O_j - X_jY_jZ_j$ is obtained as

$$W_k^I = \left[\prod_{u=n,L^n(k)=0}^{u=1} T_{L^u(k)L^{u-1}(k)}^{P} T_{L^u(k)L^{u-1}(k)}^{D}\right]^{-1} \left[\prod_{t=n,L^n(j)=0}^{t=1} T_{L^t(j)L^{t-1}(j)}^{P} T_{L^t(j)L^{t-1}(j)}^{D}\right] W_j \tag{8}$$

The actual homogeneous coordinates of the point W in subcoordinate system $O_j - X_jY_jZ_j$ is

$$W_k = \left[w_{xk}w_{yk}w_{zk}1\right]^T \left[\prod_{u=n,L^n(k)=0}^{u=1} T_{L^u(k)L^{u-1}(k)}\right]^{-1} \left[\prod_{t=n,L^n(j)=0}^{t=1} T_{L^t(j)L^{t-1}(j)}\right] \tag{9}$$

In the process of MU working (actuator moving), the error between ideal and actual motion position of point W is expressed as

$$E = \left(e_x^P \ e_y^P \ e_z^P \ 0\right)^T = W_k^I - W_k$$
$$= \left[\prod_{u=n,L^n(k)=0}^{u=1} T^P_{L^u(k)L^{u-1}(k)} T^D_{L^u(k)L^{u-1}(k)}\right]^{-1} \left[\prod_{t=n,L^n(j)=0}^{t=1} T^P_{L^u(j)L^{u-1}(j)} T^D_{L^u(j)L^{u-1}(j)}\right] W_j \quad (10)$$
$$- \left[\prod_{u=n,L^n(k)=0}^{u=1} T_{L^u(k)L^{u-1}(k)}\right]^{-1} \left[\prod_{t=n,L^n(j)=0}^{t=1} T_{L^u(j)L^{u-1}(j)}\right] W_j$$

Equation (10) is the comprehensive spatial motion error model of MU actuators for general electromechanical products.

4.2 Formation mechanism of the MU precision life

In the process of electromechanical product working, MUs are influenced by various changing indicators, e.g., increase in ambient temperature, wear on part surface, and change of load or force. These indicators cause the variation of the MU accuracy [37, 38].

The general formation process of MU accuracy faults is shown in **Figure 3**. After MU moving for a period of time, its accuracy variation reaches the designed accuracy variation limit (Y_{max} or Y_{min}). Its specific variation law $f(t)$ is as follows.

4.2.1 Initial stage

MU is working at an expectation accuracy a_0 (producing accuracy). For a specific MU, because of the influence of its initial state and working condition, a_0 has a certain dispersion, following the distribution law $f(a)$.

4.2.2 Stable stage

After MU working stably for a period of time T_a, the influence reaches enough degree to cause the variation of MU accuracy. Because of the considerable degree of

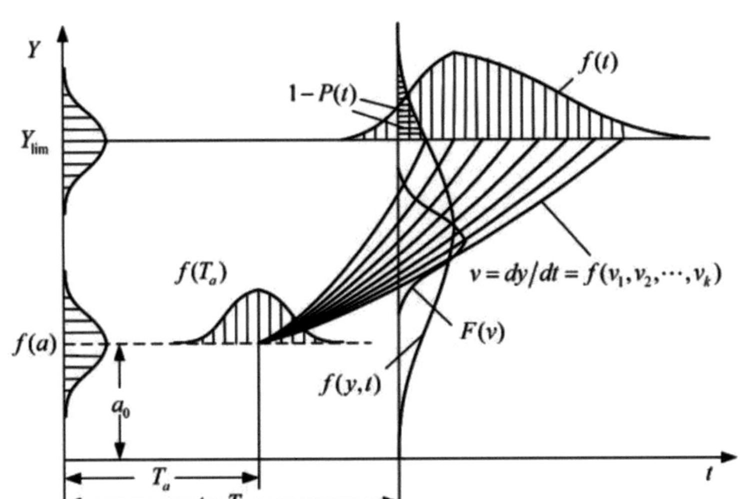

Figure 3.
The general forming process of the MU accuracy failure.

randomness of the uncertainty influencing factors, the stable operating time T_a obeys a random distribution $f(T_a)$.

4.2.3 Degradation stage

MU accuracy reaches Y_{lim} at a speed v, and faults happen, where v is a random factor obeying distribution law $F(v)$, so operating time without faults T is also a random factor obeying $f(t)$.

During the moving process, MU is affected by various factors, e.g., friction, environment, and changing working conditions. Its accuracy variation rate v can be regarded as a normal distribution:

$$f(v) = \frac{1}{\sigma_v \sqrt{2\pi}} e^{-\frac{(v-\overline{v})^2}{2\sigma_v^2}} \tag{11}$$

where $f(v)$ is the probability density function of accuracy variation, $\overline{v}$ represents the average rate of accuracy variation process, and σ_v is the mean square error of rate accuracy variation.

For a specific MU, if its initial accuracy is a known value a_0, obeying a random distribution and its stable accuracy stage is finished, that is, $T_a = 0$, and then its accuracy variation is obtained as (e.g., linear correlation)

$$Y = a_0 + vt \tag{12}$$

As is shown in **Figure 3**, before accuracy reaching the limitation, $Y = Y_{lim}$, its operating time without failure is the precision life (accuracy retention time), which is the function of speed v:

$$T = \frac{Y0_{lim}}{v} \tag{13}$$

Finally, its accuracy reliability is expressed as

$$R = P(t \le T) = P(v \le \left| \frac{Y0_{lim}}{t} \right\| \left| \frac{Y0_{lim}}{t} \right\| \Big|) = \Phi(\left| \frac{Y0_{lim}}{t \cdot \sigma_v} \right\| \frac{\overline{v}}{\sigma_v} \frac{\overline{v}}{\sigma_v} \Big\| \tag{14}$$

4.3 Formation mechanism of the MU performance stability

The performance stability of MUs refers to the ability of maintaining the performance of the MU or the variation range of output quality characteristic (dynamic characteristic) minimal with the uncertainty factor while satisfying various constraints. It means that the output value Y fluctuates around the input-output linear relationship and the smaller the fluctuation, the better the stability.

4.3.1 Performance stability evaluation of MUs

4.3.1.1 The dynamic tracking errors and steady errors

The performances of MUs are real-time and dynamic, and its output quality characteristic fluctuates with input parameters. That is, there is a corresponding objective value for each input signal. At the same time, the fluctuation needs to be as small as possible. This characteristic is defined as a dynamic characteristic.

The dynamic state in the process of MU motion is divided into transient and stead. Dynamic tracking errors mean the deviation between the actual value $y(t)$ and expected value y_{Ideal} of the MU quality characteristic indicator of each transient state. Steady errors refer to the fluctuation amplitude Δ of the deviation between the actual value $y(t > T)$ and expected value y_{Ideal} of the MU quality characteristic under steady state after moving for a time T. To guarantee enough high-performance stability, there are two premises for them:

$$\begin{cases} \delta_{max} = \left|y(t) - y_{Ideal}\right|_{max} \le \delta_{Ideal} \\ \Delta_{max} = \left|y(t>T) - y_{Ideal}\right|_{max} - \left|y(t>T) - y_{Ideal}\right|_{min} \le \delta_{Ideal} \end{cases} \tag{15}$$

4.3.1.2 The signal/noise ratio

The indicator of signal/noise ratio (SNR) [39] (Eq. 16) proposed by Prof. Taguchi is applied to evaluate the MU dynamic performance stability in this section. Generally, the greater the SNR value, the better the stability, which indicates that the noise is low even if the signal is strong:

$$\eta = \frac{P_S}{P_N} \tag{16}$$

where P_S and P_N represent the power of signal and noise, respectively.

For the dynamic characteristic, the objective value m is not a constant but a variable determined according to the variation of signal M output characteristic $y = \alpha + M\beta + \varepsilon$ (**Figure 4**), where ε reflects the degree of interference. Moreover, the variation of output characteristic is set as a normal distribution $N(0,\delta^2)$.

As is shown in **Figure 4**, the SNR of dynamic characteristic y is defined as

$$\eta = \frac{\beta^2}{\sigma^2} \tag{17}$$

where β^2 is the measure of signal sensitivity, and δ^2 is the fluctuation rate (deviation) of amplitude, which is used to evaluate the stability of MU.

For a specific MU quality indicator q_1, the greater the signal/noise ratio η_1, the more stable the quality indicator. Its weight (the contribution rate of this quality indicator to quality instability) is expressed as

$$\lambda_l = \frac{1}{\eta_l * \sum_{j=1}^{m} \frac{1}{\eta_j}} \tag{18}$$

4.3.2 Clearance formation mechanism of MUs

The clearance is one of the main factors causing noise of MU and influencing the performance stability of complete machine [40]. However, in real engineering

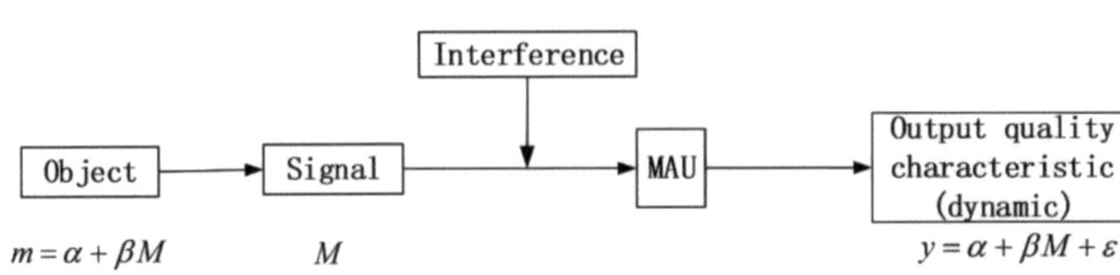

Figure 4.
The sketch of dynamic characteristic.

application, the MU clearance is inevitable. It happens in the inner composition parts of MUs, e.g., the clearance between the large piston and the inner wall of the mandrel in the piston moving unit and the meshing clearance between the rack translating unit and the gear rotating unit (**Figure 5**).

The clearance between MUs includes the static clearance of geometric size and the dynamic clearance under a moving state. As is shown in **Figure 5**, they are mainly caused by three aspects:

1. A regular clearance reserved due to the motion fit design of the MU.

2. The uncertainty factors in the MU design and manufacturing process. The error will be caused to form clearance.

3. The uncertainty factors in the process of operating, e.g., vibration, friction, wear, etc., cause the irregular clearance.

4.4 Formation mechanism of the MU reliability

The MU reliability refers to the ability to accomplish specific machine motion precisely, timely, and coordinately under a specified time and condition and maintain all quality characteristics within the allowable range. Its quantification indicator, the probability, is known as reliability.

4.4.1 Reliability indicators of MUs

4.4.1.1 The MU reliability

The MU reliability refers to the probability of output quality characteristic parameters being in the allowable range at a specified time (period). Suppose that output characteristic parameter, e.g., accuracy, precision life, performance stability, etc., is random variable $Y(t)$. According to the design requirement, output quality characteristic parameter is controlled within the range of $[Y_{min}, Y_{max}]$. The MU reliability is defined as the probability P that happens:

$$R = P[Ymax_{min}] \tag{19}$$

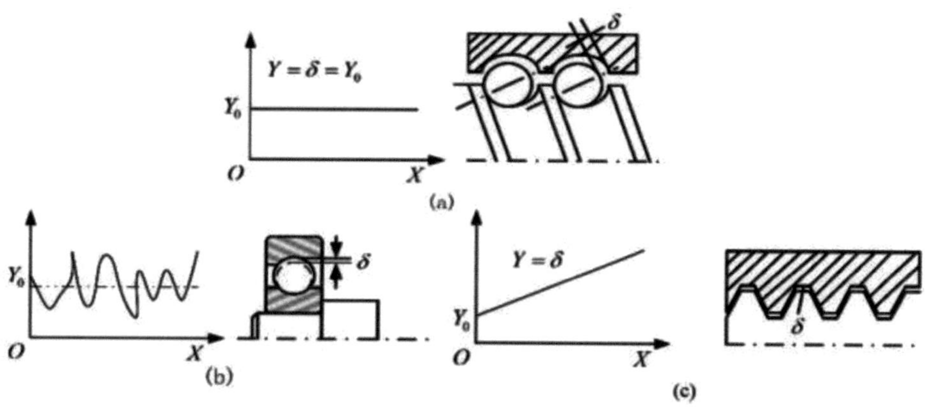

Figure 5.
Clearances between MUs resulting from (a) assembly, (b) design, and (c) wear.

Its corresponding failure probability F is expressed as

$$F = 1 - R = 1 - P[Ymax_{min}] \tag{20}$$

Taking the motion accuracy (motion error) of the MU as an example, its motion accuracy reliability is the probability of its motion output error within the maximum allowable error range:

$$R = P[emax_{min}] \tag{21}$$

For a specific MU, its error obeys a normal distribution, and its reliability is derived as

$$\begin{aligned} R &= P(e_{min} \le E \le e_{max}) \\ &= P(E \le e_{max}) - P(E \le e_{min}) \\ &= \Phi\left(\frac{e_{max} - \mu}{\sigma}\right) - \Phi\left(\frac{e_{min} - \mu}{\sigma}\right) \end{aligned} \tag{22}$$

4.4.1.2 The mission reliability of complete machine system

Because multiple different missions of the upper motion units are accomplished by the input and output of multiple MUs, from the perspective of the mission reliability, the mission reliability of the complete machine system of electromechanical products is an organic combination of the MU motion reliability (**Figure 6**), which is expressed as follows:

$$R^W = \sum_{i=1}^{n} \alpha_i R^{A_i} \tag{23}$$

4.4.2 Failure mechanism of MUs

The MU failure analysis mainly includes three aspects: the failure stress, the failure mechanism, and the failure mode. To highlight the key points, we mainly analyze them from the perspective of the motion function.

4.4.2.1 The failure stress

The failure stress is the physical condition causing the MU failure, mainly including the working and environment stress. The working stress refers to the

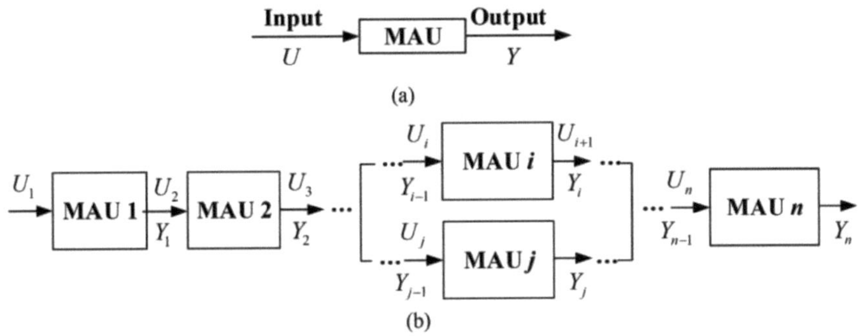

Figure 6.
The mission reliability model of (a) MU and (b) complete machine system.

necessary stress to motivate MU to achieve a basic motion function, e.g., pressure, thrust, bending torque, etc.; the environment stress refers to MU's external working environment conditions and its interaction between the internal parts, e.g., temperature, humidity, dust, noise, vibration, friction, impact force, etc.

Taking the end-toothed disk indexing turntable function unit as an example, we analyzed the failure stress of each decomposition MU (**Table 2**).

4.4.2.2 The failure mechanism

The MU quality uncertainty includes gradual and emergent uncertainty; furthermore, corresponding failures are gradual and emergent. When they happen, their certain performance parameter reaches or exceeds the threshold, caused by the MU uncertainty. If an MU performance parameter changes at the beginning of working process without a stable stage, the failure that happened is defined as a gradual failure. If the MU performance parameter suddenly reaches or exceeds the threshold in an instant with an infinite speed, the failure that happened is defined as an emergent failure. The gradual failure is also called as wear failure, mainly including:

A. Functional abnormality due to the MU wear deformation, e.g., worm gear MU rotation difficulty, loose, vibration, large clearance, etc.; B. function

Function unit	Motion unit	MU	Failure stress
End-toothed disk indexing turntable	Turntable ascending and descending	Piston moving	Hydraulic oil (hydraulic pressure, cleanliness), turntable weight, concentricity, cylindricity, burrs, clearances, assembly stress
		End-toothed disk moving up and down	Weight of the turntable and tool, cutting force and load
	Turntable rotating	Motor rotating	Temperature, cutting fluid
		Worm rotating	Temperature, load, wear, axial thrust, cyclic load stress, vibration, concentricity
		Worm gear rotating	Temperature, load, wear, shear stress
		Gear shaft rotating	Load torque, fatigue, radial load
		Upper tooth rotating	Load torque, wear
		Revolving body rotating	Cutting liquid, load
	Pallet clamping and loosening	Spring pin moving	Vibration, precision
		Spring shrinking	Load, fatigue
		Piston moving	Hydraulic oil (hydraulic pressure, cleanliness), load, concentricity, cylindricity, burrs, clearance, assembly stress
		Pull rod moving	Pull stress, vibration
		Pull claw moving up and down	Pull stress, assembly stress, fatigue, vibration, precision

Table 2.
The failure stress analysis of end-toothed disk indexing turntable function unit.

Failure reason	Failure mechanism	Cause analysis
Wear	Wear generally happens in the relative motion part of MUs. The friction factor increases gradually in the initial wear stage, accompanied by heat, which eventually causes melting	The influence factors include mainly the materials in contact with each other, the surface contact pressure, the wear rate, and the lubricating oil
Fatigue	The MU undergoes a wavelike stress during the movement, and its effect accumulates continuously and develops into a damage after a certain number of iterations	Internal force caused by vibration, rotation, intermittent motion, etc. External force due to roughness of the external contact surface
Vibration	Different degree vibrations happen during the MU motion. Some of them cause failure of the MU and are defined as "failure" vibration	Vibrations determined by the MU structural characteristics, such as meshing vibration of gears, etc.
Yielding	Deformation exceeds the elasticity limit of the material to produce unrecoverable plastic deformation	Under long time stress or stress much higher than the elasticity limit, plastic deformation is produced
Impact	When the MU structure material property is brittle, with an excessive impact load, the visible deformation, e.g., extension or bending, is caused under one shot, especially at the stress concentration	This kind of damage suddenly is caused with a large loss, which is mainly determined by the material basic properties. The reasons include the choice of materials, the stress concentration structure, the imperfect function of the buffer mechanism, etc.

Table 3.
Formation mechanisms of typical MU failures.

abnormality due to the deformation or fracture caused by fatigue, e.g., gear tooth fracture of gear rotation MU; C. deformation and fracture due to corrosion; D. deformation due to yielding; E. deformation, cracking, and fracture due to deterioration of the polymer; F. obstacles caused by wear-induced debris; G. abnormal rotation and heat caused by deterioration of oil; H. abnormal movement due to the incorporation of rust, peeling paint, and peeling plating.

The formation mechanisms of the typical MU failures are shown in **Table 3**.

5. Key quality characteristic control methods based on the MU

On the basis of studying the key quality characteristics of MUs and the coupling relationship of the complete machine quality characteristics, it is necessary to unitize and aggregate the MUs' quality characteristic according to the "decomposition-analysis-synthesis" analysis method of the large system control theory [41]. The whole machine quality prediction method is studied from bottom to top, the quality characteristic state space model is constructed, and the predictive control algorithm is derived to achieve the comprehensive prediction and control of the complete machine quality characteristics.

5.1 Multiple generalized operator model of complex large systems

The whole working process of electromechanical products is achieved and accomplished via various orderly units. Therefore, the whole control process of the complete machine quality characteristic system can be decomposed into several quality control sub-processes of units and regarded as a linear system. To better

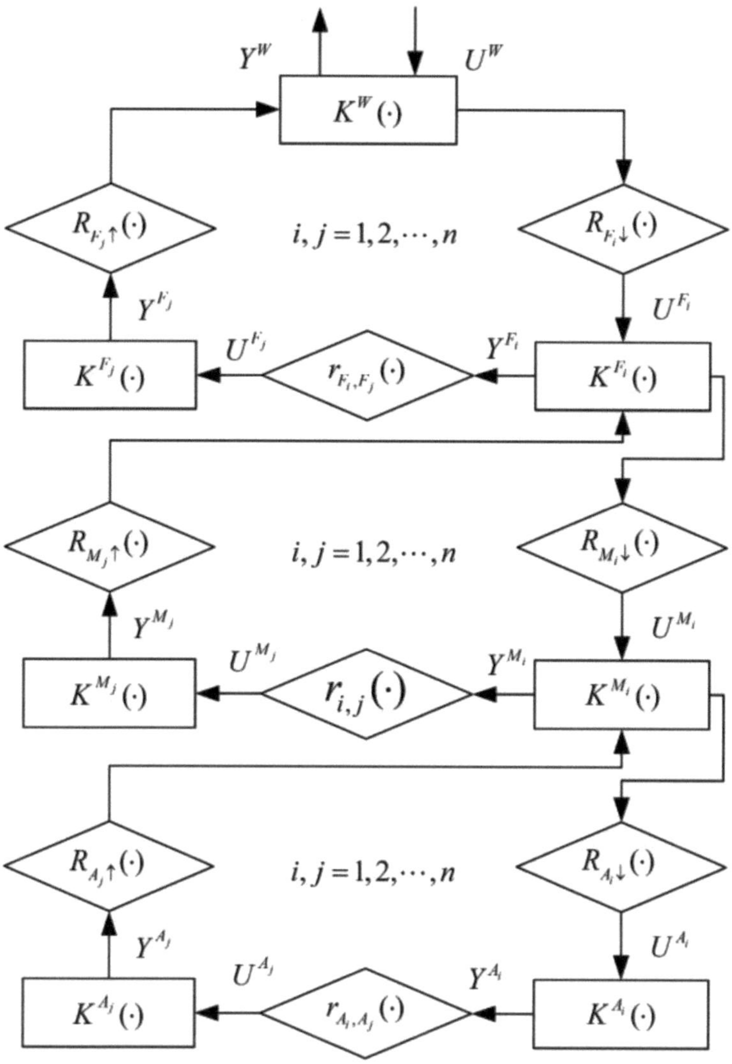

Figure 7.
The multilayer generalized operator model of complete machine.

understand the decomposition and aggregation relationship between the quality characteristics of the upper and lower units, corresponding to the quality characteristic coupling relationship model based on the FMA tree, the multiple generalized operator model of the complete machine large system [42] is established combining generalized operator model of complete machine system and the generalized operator relationship model of units (**Figure 7**).

The power input of the complete machine system is decomposed into the units' output, which is transmitted through the same layer and then aggregated to the complete system.

5.1.1 Generalized operator model of units

1. The first-layer generalized operator model of the complete machine system can be expressed as

$$Y^W \doteq K^W(\cdot)U^W \tag{24}$$

where U^W and Y^W are the total input and output of the complete machine system, respectively, and $K^W(\cdot)$ is the macro generalized operator of coarse granularity.

2. The second-layer generalized operator model of the function units can be expressed as

$$Y^{F_i} \doteq K^{F_i}(\cdot)U^{F_i}, Y^{F_j} \doteq K^{F_j}(\cdot)U^{F_j}, i, j = 1, 2, \cdots, n \tag{25}$$

where U^{F_i} and Y^{F_i} are the input and output of function unit F_i, respectively, and $K^{F_i}(\cdot)$ is its generalized operator.

3. The third-layer generalized operator model of the motion units can be expressed as

$$Y^{M_i} \doteq K^{M_i}(\cdot)U^{M_i}, Y^{M_j} \doteq K^{M_j}(\cdot)U^{M_j}, i, j = 1, 2, \cdots, n \tag{26}$$

where U^{M_i} and Y^{M_i} are the input and output of motion unit M_i respectively, and $K^{M_i}(\cdot)$is its generalized operator.

4. The fourth generalized operator model of the MUs can be expressed as

$$Y^{A_i} \doteq K^{A_i}(\cdot)U^{A_i}, Y^{A_j} \doteq K^{A_j}(\cdot)U^{A_j}, i, j = 1, 2, \cdots, n \tag{27}$$

where U^{A_i} and Y^{A_i} are the input and output of MU A_i and $K^{A_i}(\cdot)$ is its generalized operator.

5.1.2 Relationship model of generalized operators

5.1.2.1 Relationship model of vertical generalized operators

a. According to the function and mission of the unit, the total input of the complete machine system is distributed down to the corresponding unit of the next layer to form its input. Therefore, the input between units from different layers is the effective allocation of the complete machine system input, and the downward vertical relationship model can be expressed as

$$\begin{cases} U^{F_i} \doteq R_{F_i\downarrow}(\cdot)U^W, i = 1, 2, \cdots, n \\ U^{M_i} \doteq R_{M_i\downarrow}(\cdot)U^{F_i}, i = 1, 2, \cdots, n \\ U^{A_i} \doteq R_{A_i\downarrow}(\cdot)U^{M_i}, i = 1, 2, \cdots, n \end{cases} \tag{28}$$

where $R_\downarrow(\cdot)$ is the downward vertical relationship operator.

b. The output (quality characteristics) of the unit is aggregated to the complete machine system to form its total output (quality characteristics). Therefore, Y^W is the order integration of the unit output, and the upward vertical relationship model can be expressed as

$$\begin{cases} Y^W \doteq R_{F_j\uparrow}(\cdot)Y^{F_j}, j = 1, 2, \cdots, n \\ Y^{F_i} \doteq R_{M_j\uparrow}(\cdot)Y^{M_j}, j = 1, 2, \cdots, n \\ Y^{M_i} \doteq R_{A_j\uparrow}(\cdot)Y^{A_j}, j = 1, 2, \cdots, n \end{cases} \tag{29}$$

where $R_\uparrow(\cdot)$ is the upward vertical relationship operator.

5.1.2.2 Relationship model of horizontal generalized operators

$$\begin{cases} U^{F_j} \doteq r_{F_i,F_j}(\cdot)Y^{F_i}, i,j = 1,2,\cdots,n \\ U^{M_j} \doteq r_{M_i,M_j}(\cdot)Y^{M_i}, i,j = 1,2,\cdots,n \\ U^{A_j} \doteq r_{A_i,A_j}(\cdot)Y^{A_i}, i,j = 1,2,\cdots,n \end{cases} \tag{30}$$

where $r_{A_i,A_j}(\cdot)$ is the horizontal relationship operator of MU A_i and A_j.

5.2 Multilayer state space model of quality characteristics

According to the quality characteristic mapping model based on the FMA tree, the large system of the complete electromechanical product is decomposed into several units. To predict and control the quality characteristics of different hierarchical units, the corresponding granularity required is different. The structural features of the multilayer state space model system, decomposition-analysis-synthesis, can combine the microscopic quality characteristics of each unit and the macroscopic large system to control the quality characteristics more accurately. Therefore, referring to the multigeneralized operator model of the large-scale system, the multilayer state space model of the quality characteristic of the complete machine is established step by step as follows [43, 44].

5.2.1 Decomposition

According to the qualitative model, the quantitative equations of multilayer state space model of the complete machine quality characteristics are set as follows:

$$M_I : \begin{cases} \dot{x}_I^W = A_I^W x_I^W + B_I^W u_I^W \\ y_I^W = C_I^W x_I^W + D_I^W u_I^W \end{cases} \quad M_{\text{II}}^{F_i} : \begin{cases} \dot{x}_{\text{II}}^{F_i} = A_{\text{II}}^{F_i} x_{\text{II}}^{F_i} + B_{\text{II}}^{F_i} u_{\text{II}}^{F_i} \\ y_{\text{II}}^{F_i} = C_{\text{II}}^{F_i} x_{\text{II}}^{F_i} + D_{\text{II}}^{F_i} u_{\text{II}}^{F_i} \end{cases}$$

$$M_{\text{II}}^{F_j} : \begin{cases} \dot{x}_{\text{II}}^{F_j} = A_{\text{II}}^{F_j} x_{\text{II}}^{F_j} + B_{\text{II}}^{F_j} u_{\text{II}}^{F_j} \\ y_{\text{II}}^{F_j} = C_{\text{II}}^{F_j} x_{\text{II}}^{F_j} + D_{\text{II}}^{F_j} u_{\text{II}}^{F_j} \end{cases} \quad M_{\text{III}}^{M_i} : \begin{cases} \dot{x}_{\text{III}}^{M_i} = A_{\text{III}}^{M_i} x_{\text{III}}^{M_i} + B_{\text{III}}^{M_i} u_{\text{III}}^{M_i} \\ y_{\text{III}}^{M_i} = C_{\text{III}}^{M_i} x_{\text{III}}^{M_i} + D_{\text{III}}^{M_i} u_{\text{III}}^{M_i} \end{cases},$$

$$M_{\text{III}}^{M_j} : \begin{cases} \dot{x}_{\text{III}}^{M_j} = A_{\text{III}}^{M_j} x_{\text{III}}^{M_j} + B_{\text{III}}^{M_j} u_{\text{III}}^{M_j} \\ y_{\text{III}}^{M_j} = C_{\text{III}}^{M_j} x_{\text{III}}^{M_j} + D_{\text{III}}^{M_j} u_{\text{III}}^{M_j} \end{cases} \quad M_{\text{IV}}^{A_i} : \begin{cases} \dot{x}_{\text{IV}}^{A_i} = A_{\text{IV}}^{A_i} x_{\text{IV}}^{A_i} + B_{\text{IV}}^{A_i} u_{\text{IV}}^{A_i} \\ y_{\text{IV}}^{A_i} = C_{\text{IV}}^{A_i} x_{\text{IV}}^{A_i} + D_{\text{IV}}^{A_i} u_{\text{IV}}^{A_i} \end{cases}$$

$$M_{\text{IV}}^{A_j} : \begin{cases} \dot{x}_{\text{IV}}^{A_j} = A_{\text{IV}}^{A_j} x_{\text{IV}}^{A_j} + B_{\text{IV}}^{A_j} u_{\text{IV}}^{A_j} \\ y_{\text{IV}}^{A_j} = C_{\text{IV}}^{A_j} x_{\text{IV}}^{A_j} + D_{\text{IV}}^{A_j} u_{\text{IV}}^{A_j} \end{cases}$$

where $y(t)$ is the output variable, $y = (y_1, y_2, \cdots, y_m)^T$; $x(t)$ is the state variable, $x = (x_1, x_2, \cdots, x_n)^T$; and $u(t)$ is the input variable, $u = (u_1, u_2, \cdots, u_r)^T$. A is the object (system) matrix describing the characteristics of the controlled object:

$$A = \left[a_{ij}\right]_{n \times n} = \begin{bmatrix} a_{11} & a_{12} & \cdots & a_{1n} \\ a_{21} & a_{22} & \cdots & a_{2n} \\ \vdots & \vdots & \ddots & \vdots \\ a_{n1} & a_{n2} & \cdots & a_{nn} \end{bmatrix} \tag{31}$$

B is the control matrix describing the characteristics of the control mechanism:

$$B = \left[b_{ij}\right]_{n\times r} = \begin{bmatrix} b_{11} & b_{12} & \cdots & b_{1r} \\ b_{21} & b_{22} & \cdots & b_{2r} \\ \vdots & \vdots & \ddots & \vdots \\ b_{n1} & b_{n2} & \cdots & b_{nr} \end{bmatrix} \tag{32}$$

C is the observing matrix describing the characteristics of the observing device:

$$C = \left[c_{ij}\right]_{m\times n} = \begin{bmatrix} c_{11} & c_{12} & \cdots & c_{1n} \\ c_{21} & c_{22} & \cdots & c_{2n} \\ \vdots & \vdots & \ddots & \vdots \\ c_{m1} & c_{m2} & \cdots & c_{mn} \end{bmatrix} \tag{33}$$

D is the observing matrix describing the characteristics of the observing device:

$$D = \left[d_{ij}\right]_{m\times r} = \begin{bmatrix} d_{11} & d_{12} & \cdots & d_{1r} \\ d_{21} & d_{22} & \cdots & d_{2r} \\ \vdots & \vdots & \ddots & \vdots \\ d_{m1} & d_{m2} & \cdots & d_{mr} \end{bmatrix} \tag{34}$$

5.2.2 Synthesis

Because the complete electromechanical products are composed of a series of units, according to the research on coupling and decoupling technology based on MUs, the spatial state variables of the complete product can be expressed as the weighted sum of that of some function units. Furthermore, the spatial state variables of function units can be expressed as the weighted sum of that of some motion units. Finally, the spatial state variables of motion units can be expressed as the weighted sum of that of some MUs. The specific model can be expressed as

$$\begin{cases} x_I^W = w_i x_{\text{II}}^{F_i} + w_j x_{\text{II}}^{F_j} \\ x_{\text{II}}^{F_i} = w_i x_{\text{III}}^{M_i} + w_j x_{\text{III}}^{M_j} \\ x_{\text{III}}^{M_i} = w_i x_{\text{IV}}^{A_i} + w_j x_{\text{IV}}^{A_j} \end{cases} \tag{35}$$

6. Discussion

With the increase of structure complexity of electromechanical products, the quality control of electromechanical products by parts is too complicated, and its quality control becomes significantly difficult. The control method based on MUs provides a good idea properly analyzing the formation mechanism of the key quality characteristics (precision, precision life, performance stability, and reliability) of electromechanical products is studied based on MUs. On the one hand, due to the appropriate granularity, it is more convenient for quality control; on the other hand, because the meta-action decomposition method is a dynamic decomposition process of electromechanical products, the quality characteristics of electromechanical products can be controlled by the four key quality characteristics (the precision, the precision life, the stability, the reliability) of the quality of the control MUs. It makes up for the deficiency of traditional control methods.

7. Conclusions and prospects

The research shows that it is adaptive and feasible to study the formation mechanism of key quality characteristics and quality control through the proposed analysis method based on the meta-action theory. MU has specific functions and can independently complete the specified motion or operation, which is more suitable for quality and reliability analysis. In this paper, the forming mechanism of MU's four key quality characteristics (the accuracy, the precision life, the performance stability, the reliability) is deeply studied, and its parameter expression model is obtained. The model of quality characteristic state space of unit is established; based on that, the basic principle of predictive control and the forming mechanism of the quality characteristics of MUs are analyzed and clarified. Starting from the quality control of MUs, the quality control of the whole machine is accomplished through the quality control of the unit level. This control mode is more simple, effective, and conducive to the accomplishment of refined quality control.

Due to a variety of electromechanical products, different functions and their implementation processes are complex. On the one hand, the data acquisition of MU quality index is very difficult. On the other hand, the factors influencing the quality uncertainty are various. Moreover, the meta-action theory also has certain limitations up to now, and some deficiencies exist here. Thus, we plan to further improve the following research work in the future:

1. In the follow-up study, the quality characteristic database of meta-action units should be continuously increased through various channels such as collecting process data or conducting meta-action unit tests so as to better verify the theoretical methods.

2. This paper only studies the four key quality characteristics (the reliability, the precision, the precision life, the performance stability). In order to make the later quality prediction and control more accurate, it can be considered to increase the research of quality characteristics such as usability.

Acknowledgements

This work is financially supported by the National Natural Science Foundation of China (No. 51705048; 51835001) and the National Major Scientific and Technological Special Project for "High-grade CNC and Basic Manufacturing Equipment" of China (2018ZX04032-001; 2016ZX04004-005).

Author details

Yan Ran, Xinlong Li, Shengyong Zhang and Genbao Zhang
Chongqing University, China

*Address all correspondence to: ranyan@cqu.edu.cn

References

[1] Yan Z, Liuyong S, Liu Shihao HS. Exploration and practice of electromechanical integration talent training mode based on social needs. Natural Science Journal of Hainan University. 2014;**32**:389-393. DOI: 10.15886/j.cnki.hdxbzkb.2014.04.017

[2] Feng Z. Study on Organizational Structure Optimization Design of Xi'an Dongfeng Instrument and Meters Factory. Xian: Northwest University; 2012

[3] Zhang G, Jihong P, Ren Xianlin CHG. Research on reliability analysis technology of typical meta-action units of NC machine tools. Computer Integrated Manufacturing Systems. 2011;**17**:151-158

[4] Mengsheng Y. Research on Reliability Analysis Technology of Typical Meta-Action Units of NC Machine Tools. Chongqing: Chongqing University; 2018

[5] Sata T, Takeuchi Y, Okubo N. Improvement of working accuracy of a machining center by computer control compensation. In: Tobias SA, editor. Proceedings of the Seventeenth International Machine Tool Design and Research Conference; 20–24 September 1976; Birmingham. London: Macmillan Education UK; 1977. pp. 93-99 10.1007/978-1-349-81484-8_12

[6] Ceglarek D, Shi J, SM W. Fixture failure diagnosis for autobody assembly using pattern recognition. ASME Journal of Engineering for Industry. 1996;**118**: 55-66. DOI: 10.1115/1.2803648

[7] Zeyuan D, Li J, Xinjun L, Mei Bin CJ. Experimental study on the effectiveness of two different geometric error modeling methods for machine tools. Journal of Mechanical Engineering. 2019;**55**:137-147. DOI: 10.3901/JME.2019.05.137

[8] Li J, Fugui X, Xinjun L, Mei Bin DZ. Analysis on the research status of volumetric positioning accuracy improvement methods for five-axis NC machine tools. Journal of Mechanical Engineering. 2017;**53**:113-128. DOI: 10.3901/JME.2017.07.113

[9] Yongwei Y, Liuqing D, Yi Xiaobo CG. Prediction method of NC machine tools' motion precision based on sequential deep learning. Transactions of the Chinese Society of Agricultural Engineering. 2019;**50**: 421-426. DOI: 10.6041/j.issn.1000-1298.2019.01.049

[10] Min W, Rui S, Wei Z, Deshun K, Shuang Z. Accelerated degradation test method for accuracy stability of precision ball screws. Journal of Beijing University of Technology. 2016;**42**: 1629-1633. DOI: 10.6041/j.issn.1000-1298.2019.01.049

[11] Liping Z, Yenong L, Yang Q. Gray prediction about of the spindle rotation accuracy of CNC lathe. Machine Tool and Hydraulics. 2016;**44**:93-97. DOI: 10.3969/j.issn.1001–3881.2016.21.022

[12] Linlin Y. Study on Accuracy Stability Testing of Servo Feed System on CNC Machine tool. Hangzhou: Zhe jiang University; 2014

[13] Hu M, Yu CW, Zhang J, Zhao W, Cun H, Yuan S. Accuracy stability for large machine tool body. Journal of Xi'an Jiaotong University. 2014;**48**: 65-73. DOI: 10.7652/xjtuxb201406012

[14] Kang XM, Fu WP, Wang DC, Li T, Wang SJ. Analysis and testing of axial load effects on ball Screw's friction torque fluctuations. Noise and Vibration Control. 2010;**30**:57-61. DOI: 10.3969/j.issn.1006-1355.2010.02.057

[15] Saitou K, Izui K, Nishiwaki S, et al. A survey of structural optimization in

mechanical product development. Journal of Computing & Information Science in Engineering. 2005;**5**(3): 214-226. DOI: 10.1115/1.2013290

[16] Ta TN, Hwang YL, Horng JH. The influence of friction force on sliding stability of controlled multibody systems—CNC machine tool. Jurnal Tribologi. 2018;**19**:107-120

[17] Caro S, Bennis F, Wenger P. Tolerance synthesis of mechanisms: A robust design approach. 2010;**127**(1): 339-348

[18] Gao YC, Feng YX, Tan JR. Product quality characteristics robust optimization design based on minimum sensitivity region estimation. Computer Integrated Manufacturing Systems. 2010;**16**:897-904

[19] Hui L, Huang Y, Huijie ZH. Effects of transmission stiffness variations on the dynamic accuracy consistency of CNC feed drive systems. Journal of Mechanical Engineering. 2014;**23**: 128-133. DOI: 10.3901/JME.2014.23.128

[20] Zhang X, Gao H, Huang H-Z, Li Y-F, Mi J. Dynamic reliability modeling for system analysis under complex load. Reliability Engineering and System Safety. 2018;**180**:345-351. DOI: 10.1016/j.ress.2018.07.025

[21] Pinghua J, Guangquan H, Yan R, Xiao Liming LZ. Reliability control method of assembly process of products based on motion unit fault model. Journal of Central South University (Science Technology). 2018;**9**: 2197-2205. DOI: 10.11817/j.issn.1672-7207.2018.09.012

[22] Zhang G, Liu J, Ge H. Modeling and analysis for assembly reliability based on dynamic Bayesian networks. Chinese Journal of Mechanical Engineering. 2012;**23**:211-215. DOI: 10.3969/j.issn.1004-132X.2012.02.019

[23] Assaf T, Dugan JB. Diagnosis based on reliability analysis using monitors and sensors. Reliability Engineering and System Safety. 2008;**93**:509-521. DOI: 10.1016 /j.ress.2006.10.024

[24] Yu FJ, Ke YL, Ying Z. Decision on failure maintenance for aircraft automatic join-assembly system. Computer Integrated Manufacturing Systems. 2009;**15**:1823-1830

[25] Qinghu Z. Fault Prognostics Technologies Research for Key Components of Mechanical Power and Transmission Systems. Changsha: National University of Defense Technology; 2010

[26] Haifeng Z. Coupling and Decoupling Control of Complex Product Quality Characteristics. Chongqing: Chongqing University; 2010

[27] Xianghua A. The Theory and Application of Multi-Scale Intelligent and Cooperative Quality Control for Key Parts of Large Air Separation Equipment. China: Zhejiang University; 2011

[28] Nada OAA. Quality Prediction in Manufacturing System Design. Canada: University of Windsor; 2016

[29] Xianlin R. Research on the Key Technology of Quality Characteristic Prevention and Control of Electromechanical Products. China: Chingqing University; 2011

[30] Zhentao S. Research of Coupling and Prediction Control of Multiple Quality Characteristics. University of Electronic Science and Technology of China; 2016

[31] Yang JP, Wang WL, Kang J, Mi SF. Research on multi-phased product quality predictive control method based on PSO-SVM. Journal of Dalian Nationalities University. 2013;**15**:37-41.

DOI: 10.3969/j.issn.1009-315X.2013.01.009

[32] He Z, Zou Feng ZY. Study and application on the integration of quality Emerging Trends in Mechatronics tools based on QFD. Modul Mach TOOL Modular Machine Tool & Automatic Manufacturing Technique. 2006:196-199, 102. DOI: 10.3969/j.issn.1001-2265.2006. 04 01.031

[33] Houston RL. Multibody System Dynamics. Vol. 1. Tianjing: Tianjin University Press; 1987

[34] Yangmin L. Kinematics analysis of multibody system of loader working mechanism. Journal of South China University of Technology. 1996;**24**(2): 84-91

[35] Jinwei F. Establishing spatial error model of CNC machine tool by using multibody system kinematics theory. Journal of Beijing University of Technology. 1999;**25**(2):38-44

[36] Shiping L. In: National University of Defense Technology, editor. Research on Precision Modeling and Error Compensation Method for Multi-Axis CNC Machine Tools. Changsha; 2002

[37] Li C, Yu Lijiyi PA. Precision and Reliability of CNC Machine Tools. Mechanical Industry Press; 1987

[38] Hegadekatte V, Huber N, Kraft O. Finite element based simulation of dry sliding wear. Modelling and Simulation in Materials Science and Engineering. 2005;**13**(1):57-75

[39] Taguchi G, Elsayed EA, Hsiang TC. Quality Engineering in Production System. New York: McGraw Hill; 1986

[40] Tasora A, Prati E, Silvestri M. Experimental investigation of clearance effects in a revolute joint. In: Proceedings of the AIMETA International Tribology Conference; Rome, Italy: 2004. pp. 14-17

[41] Xuyan T, Wang Cong GY. Large System Control Theory. Beijing: Beijing University of Posts and Telecommunications Press; 2005

[42] Liu Dianting LM. Study on multi-layer generalized operator model of green product design and its optimization under uncertainty. Machinery Design & Manufacture. 2014;**5**:267-269, 272. DOI: 10.3969/j.issn.1001-3997.2014.05.081

[43] Mokeev AV. Description of the digital filter by the state space method. International Siberian Conference on Control & Communications IEEE. 2009. DOI:10.1109/SIBCON.2009.5044842

[44] Aiyan W, Guangping Z. Research of multi-layer intelligent modeling method. Computer Science. 2014;**41**: 253-257, 285. DOI: 10.3969/j.issn.1002-137X.2014.03.054